The Inquisitive Pioneer vol.II

The book of At-Home Basic-Materials Electricity & Magnetism Science Activities solving with a Slide Rule

Bryan Purcell

© 2015

This is for my stellar electromagnetic left-hip, Linda W.

Introduction

This book begins with many of the key ideas from the first book, The Inquisitive Pioneer — which is based on my web site www.cosmicquestthinker.com . This is a book for a thinking person wanting to take a personal journey into fundamental science principles involving electricity through investigation. The primary person it is for is one who is inquisitive, one who not only likes science and math but prefers that science is explored in a hands-on method using mostly basic, even everyday materials in addition to needed electrical, magnetic, and electronic materials one may have to obtain to undertake these Activities.

The concentration of this book, The Inquisitive Pioneer vol II is to focus on the particular world of Electricity and Magnetism. Why? This is easy to answer due to the fact that electrical energy is the heart of all levels of technology for over a century since its discovery. From the day people such as Ben Franklin, Michael Faraday, Volta, Tesla, Thomas Edison, and a host of others worked with electricity and its applications, our lives and our world have never been the same. Electricity is the ultimate tool that allows light at night, cars to operate, heating and cooling, and presently powers the computers we use and rely upon. Think too for a moment of all the places items like electric motors touch : washers, dryers (clothing and handheld), elevators, escalators, drills, pumps, fans, air conditioners, turbines, cars, and many more than one could readily list. Clearly then a book about the fundamentals of electricity and its closely related kin magnetism is essential to uncovering science in general and to be in-step with today.

Consider electrical power for a moment — where does it come from? This is an ongoing problem being looked at the world over in the present, as the world's population increases and the demands for electricity grow exponentially. Many voices are calling for new sources, such as the renewable — wind and solar while some see the need for more nuclear plants. Even if we use dams or burn coal, all of these sources are doing one thing — generating electrical potential and that is what is essential.

This book will not answer the question of where to look or why. These are questions we will be working with for this century and beyond the world over since this power will be used not only for lighting, heating & cooling, and communication & data storage and transmission today but also transportation, purification of water, cleaning our air, and many other as of yet thought of uses for ubiquitous electricity.

This book has as its goal to examine ideas about power generation, the basics of electricity and magnetism, how much we each use in our homes, and even explore some unique electrical appliances to uncover other natural phenomena in science such as using an electric travel hot-water pot to find the specific heat capacity of water and in another case use a microwave to find the speed of light! It will make you generate questions, provide ideas, and have many paths of learning about electricity and magnetism.

What each of these Activities takes is the science-minded person. The science-minded person views the world from the scientific method approach. There is the phenomena or question, which sparks one's curiosity to do some research and reading. This then leads to posing a measurable hypothesis. From there the science-minded person gathers materials that can readily be at hand or are easy to obtain for use and be used to reach the goals set forth in the hypothesis. Experiments are ran, measurements taken, and numbers analyzed via mathematics to reach conclusions. All along it is the mind of the student of life that assembles these thoughts, these ideas, these numbers measured and mentally constructs a model of what is.

Science is the story of what is, and is the fabric woven through the whole of the universe and ourselves. We as scientists, uncover the story of science through qualitative descriptions utilizing science nomenclature and the quantitative depictions from the language of mathematics.

We as teachers, parents, and even as students undertaking the journey of science need to emphasize it is our mind's ability to take in what is, organize the information, and decipher the story of the fabric of the whole of the cosmos from the subatomic to the galactic.

In essence, science is learned by doing. It is useful to have good historical stories of the story of science and the history of those who endeavored before us and see what is, but it is also useful to plant seeds to allow a platform to speak from, a place to explore and experiment and refine our knowledge as necessary.

The Science activities in this book are called hands-on, but many books say this. Here the science explorations go from observing to actually doing the science. All activities have as their core the need to have controlled experiments with measurements and the analysis of the data. This is the heart of true science. Observational and qualitative science are the beginning positions for many things, but this book goes the extra step to all things measured, all measurements used for reasoning and reaching conclusions. We become the researcher, the experimenter, and the mathematician.

The philosophy of the book in these Science activities is to encourage an independent thinker and learner. Find the pace you want from the activities. Many activities have a prelude that gives some of the historical and/or mathematical foundations of the exploration. When encountering new ideas, read and reflect on them as well as doing further research. When approaching the activity – read it through and envision it and what should happen first. In setting it up, test it and watch what happens even before measurements. Be willing to redo experiments as needed. With measurements, watch the data, its trends, make mental predictions, estimate the answers sought and the conclusions your experiment is pointing to. The majority of materials in the lists are mostly low in cost – they can mostly be found at home right from the start – such as rulers, measuring tapes, tape, measuring cups, string and others perhaps. If not, then they are readily found in inexpensive resource places such as dollar and dollar & more sort of stores. In some cases there are a few recommended items that can readily be found online with Amazon and the like, and are often low in cost as well, such as a ceramic magnets, certain size resistors or capacitors, a small electric model motor, alligator clip wires, batteries, LEDs, and a good multimeter.

Think, too, to the history of science and math – most of it was forged not only in times of only basic materials but more importantly in the minds of those who gleaned it from nature itself by creating analogies between basic materials used in the lab to explore ideas and phenomena on all scales right outside the lab doors. Here too we as readers of this book and the doers of the science become the inquisitive pioneers like the scientists and mathematicians of yesteryear.

There is a basic reasoning to this book. Besides making science and math easy to do, it brings it to the everyday and brings it home. It allows science to jump off the page of a text book, allows it to leave the classroom, and take it home and have it spring forth from the mind of the scientifically-minded student. The use of basic materials is to help to create tangible models to think about, recall and learn from in order to grasp key ideas in science. This reasoning also to applies to the use the slide rule as the math tool to prompt the mind to take in the data, the numbers, and use them in the relations between the variables as found in the equations and uncover the relations that are indeed found in nature.

Why then, the Slide Rule as the analytical tool of the data you measure ?

This is answered in my opening chapter is my original essay examining the idea of using a slide rule in the classroom of today (see ch. I) – despite all of the available technology. The primary reason is concluded below : conclusions and goals of the book are quite simple :

The slide rule, the tangible bridge for the student, can connect all the core classes and is connecting math to the world of science, carries with it history, famous people and places, and promotes active math engagement by the student. This is used in conjunction with the rationale of the opening of the book where science is tangible and can readily be extended from basic materials which promotes models in the mind to understand concepts and which are explored through carefully crafted experiments by being the Inquisitive Pioneer! Resolve, Solve, Evolve! Explore and Enjoy! – Bryan Purcell

Table of Contents

Table of Contents

Important Note of Responsibility :

In the case of all of these Activities the following is to be adhered to – All children and students – you need the support, permission, and help from parents, guardians, and/or teachers to do these Activities. Parents and children alike – read ahead through the whole of the Activity so as to anticipate where there may be areas of concern and important levels of awareness. Always employ safe practices when using any items, such as wearing goggles. Be intelligent, mature, and cautious in the use of items – these are tools and you are using them for specific ends in order to learn. Be aware of all concerns – such as the use of hot or cold items, avoid foods that may have allergies associated with them (nuts, et al), sharp objects (knives, scissors, and other edges) – and act appropriately, make smart decisions, and be safe. When done with Activities always put away materials, clean up the area, and then address the calculations portion of the Activity.

Master List of Items
This is a basic list of nearly all the items needed for all of the Activities – with the exception of the particular electronic items (such as resistors, capacitors, bread board for circuits and the like). Recognize that any one Activity will require only a small handful of these and in most cases you might find alternatives that can act as good substitutions. Most items can be found at Dollar and Dollar and More type stores along with hardware stores.

Slide Rule	Stopwatch	Paper
Ruler	Watch or Clock	Graph Paper
Meter or Yard Stick(s)	Styrofoam Plates	Pencil / Pen / Markers
Measuring Tape	Desk & Table Lamp	String (kite)
Sewing Measuring Tape	Styrofoam Peanuts	Styrofoam & Plastic Cups
Measuring Cup Set	Penlight Laser	Styrofoam sheets
Kitchen Mass Scale	Diffraction Grating	Nuts & Bolts
Compass (N, S, E, W)	Ceramic Magnet	Plastic Wrap
Multimeter	Magnet Wire	Aluminum Foil
Paper Towel	Scissors	Paper Clips
Thermometers (lab quality)	Small Model Electric Motor	Light Bulbs (various wattages)
Travel Coffee Pot	Stack of Books	Lemons, Potatoes, Bananas
Rubber Bands	Solar Cell(s)	Plastic sandwich bags
Handheld Hair Dryer	Stereo w/ microphone	Water
Tension Scale	Wires with alligator clips	Salt
Protractor	Balloon	Large Pop Bottle (2 L)
Goggles	Tape (regular & duct)	Plastic Crate
Hole Punch	Iron Nails	String
35 mm Film Canister	Diode (various)	Outdoor Items : Sun
Resistors (various)	Capacitors (various)	LEDs (various)
Use of Indoor Items : Refrigerator, Freezer, Stove, Table, Microwave	Coins : dimes, nickels, pennies, quarters	Batteries (AA, AAA, C, D, 9V)

Ch.I
Why bring the Slide Rule back to the Classroom today?

Using Old Tools to Solve a Modern Problem
The Reintroduction of the Slide Rule to the Clasroom
Bryan Purcell

Originally adapted from my article in
Oughtred Society Journal Volume 19, Number 1, Winter, 2010

In the world of education today there is a greater emphasis on trying to find the best way to create new paths to success in learning. Most of these emphasize not only greater capacity in one's knowledge base but also the employment of technology to achieve these ends. The argument goes that technology is the best inroad since it already exists. Why? Basically the quick conclusion comes from these points : children today were born into this era of technology hence have familiarity with it, and technology is the backbone of tools in use today in the workplace.

Though these seem to be valid arguments, they overlook the most critical part of education : the process of learning is not the end product. For example these tools enable a quick solution to a story problem, but the answer is not the goal of education. It is instead the acquisition of skills to enable problem solving and the employment of these skills. This is the fundamental goal of both science and math education today. It is parallel to saying that results speak for themselves in the scientific process, but it is the assessment of the student's good data acquisition and analysis that is the most important.

Because of this, the original assertion by education today, that the use of technology is a necessity, can be considered potentially invalid and there may be other paths to the goal. One of these paths is the use of the slide rule as a math tool, not only to solve problems but also to act as a visual bridge to force the user to engage her mind, tie the concepts of the ideas, the data, and the formulae to the real world and come up with a reasonable answer to the question at hand.

The first area to examine in this dialogue is why such an interest in math and science education? The answer is obvious – the majority of the available jobs, even in a tense economy, are in the areas of math and science – such as jobs in medical fields, engineering, computer technology, and the like. A National Science Board report of 2008 mentions that the needs in these sectors will be triple that of the rest of the job market. Also, the greater one's level of education, the greater one's chances are to find employment as well as to have lifelong higher career earnings. It could even be argued that one's level of opportunity and flexibility in the market and what level to which one ascends are directly related to one's math and science skills.

Compare for example the math skills needed today to operate point-of-service cash registers that have pictures of hamburgers and fries as compared to the real powerbrokers in a business, not the CEO, but the CFO (Chief Financial Officer). These same parallels exist in the realm of investigatory science and research such as found in entry-level assistants and chief engineers.

America's greatest competitive edge has always been found in its creative efforts in the areas of science and math, which launched many endeavors such as the NASA golden decade of the 1960s to go from leaving the planet, stepping into space, and safely undertaking the greatest mission in humanity's history, the journey to the moon.

A second reason for the importance of math and science education is the fact that there is a great deal of academic competition that is not only national but international in scope, and often the skills of American students have been shown to be lacking in math and science. Numerous reports have shown that American students are regularly taking math remedial courses in college; their scores on national tests in these areas are low, and even basic math skills are lacking (The Final Report of the National Mathematics Advisory Panel in 2008 from the U.S. Dept of Education).

To address the issue, the question then arises, "What approach works best?" The argument posed here will explore both the areas of necessary concern and the application of a solution in the form of the slide rule to act as the best tool to affect one's success. The topics to explore in brief are :

1. Estimation and Basic Math Operations,
2. Simple Formula Manipulation and Understanding,
3. Math Areas of Ratios & Fractions & Proportions & Conversions, and finally
4. the Math Topics of Significant Figures & Scientific Notation understanding and use.

The abilities of Estimation and Basic Math Skills go hand in hand. These need to start early and continue to expand as a student progresses through school. The emphasis should be on the student's acting as the computer and not the machine. Reliance on the calculator shifts responsibility from the person to the machine. The answer magically appears in the window of the machine and it does not include estimation at all. The slide rule, however, necessitates that one must practice and use basic math skills and continually employ estimation in order to answer questions.

How many teachers could tell the anecdotal story of the set of students who ask for a calculator before giving an estimate or an answer and with the machine they ask whether they should multiply or divide the numbers? A slide rule cannot be used unless one begins to master these basic skills and employ a mastery of basic numeracy.

Also realize that the slide rule is a natural extension of practices already in place in most elementary school systems. In order to teach numbers, their relative sizes, and concepts like addition and subtraction, the number line is considered the best visual tool. Unlike all of the other colorful tools which have an entertainment value, the number line yields the answer. In fact two of them places alongside each other help the process of learning addition and subtraction.

The same argument is true for the slide rule. Instead of a linear line, it is a logarithmically spaced line and is useful due to the properties of logarithms for visualizing multiplication and division easily. Notice that it would be the next logical step in education; if the number line works, why cannot the slide rule?

To carry this idea further, the next area of concern is formulae and their manipulation. Most formulae in school are linear (such as area of rectangles, miles per gallon, cost per unit item, density, average speed, force, pressure, and even Ohm's Law) and are readily found on a slide rule. An important note : National Standards have these and many more formulae for which students are accountable today.

To illustrate, take distance as a value on the C scale, set it over the time on the D scale and opposite the D index is the average speed. One scale is one variable and the other scale is the other key variable in a formula. One could easily explore relations quickly and effectively. For example, 'how much time at a given speed will it take to cover some given distance?' and the like. Notice the visual link of distance and time needed for a given speed. One has to read across the scale. Conceptualizing the changing of one variable and its effect on another is very easy having this tool.

The next area for exploration primarily is concerned with proportions and conversions. Here the slide rule wins hands down! One easily can solve proportions faster with a slide rule than one can with a calculator. Also, conversions can be treated as a proportion (as can the aforementioned 3-variable functions). Many studies, too, illustrate the lack of skill in converting decimal values to fractions and vice versa. The slide rule accomplishes this visually and shows all related fractions to a given decimal value instantly as contrasted with the ubiquitous calculator.

Finally, in the area of significant figures and scientific notation, the slide rule is again the master math tool. In the real world, we need typically no more than 2 or 3 digits of value in answer. No one measures a room's length and width, and then calculates the area to the 4th or 5th decimal place when buying carpeting or tiles. Also consider the goal here: to acquire problem solving skills. This being the case, does one really gain by multiplying a number with 5 digits with another one? How the slide rule is of value here is that the typical slide rule is accurate to 2-3 digits despite the size of the number.

This last statement is explained by scientific notation, which is of such a value and is directly related to the slide rule. The slide rule has only the numbers 1 to 10 on a typical C scale, yet in reality it has all the numbers that exist! The user must merely put the number in scientific notation. In math, the multiplication of exponents or division of the exponents is readily handled by addition and subtraction.

There are some important final thoughts on these matters where the slide rule is of great interest. First, recognize that no studies of the slide rule have ever been done, not even as compared to the calculator. In the same line the skeptic might add that it is an antique. In a parallel argument, why then do we use measuring tapes still when there are electronic devices for distance, why not just use a microwave instead of an oven and stove (how many cooking shows use the microwave over traditional oven). And finally, since we primarily use digital clocks, why then continue teaching traditional clock reading?

Ultimately in this idea is the question between the Slide Rule and the Calculator :
Which is best for improving math skills and numeracy? The argument has been presented. It is sound in reasoning and consideration. Finally in this case, to overlook a hypothesis is poor science at best.

Are there other benefits not noted to the slide rule? First, unlike the calculator, the slide rule has an extensive history which can help spark the imagination of presentation and packaging of the ideas about it and its use. It was the most powerful math tool in the history of all handheld devices for 350 years.

Second, the slide rule is directly connected to famous names such as Newton, James Watt, William Oughtred, Joseph Priestly in terms of its construction and use and to those who used it such as Einstein, Hans Bethe, and von Braun as well as including numerous mathematicians, scientists, and engineers.

Third, the slide rule was the first tool outside the human mind used to create most mobile and immobile structures in society such as the Empire State Building, the Golden Gate Bridge, the jet engine, the Panama Canal, and even the Apollo spacecraft.

Third, there are a number of websites that illustrate how to use the slide rule (some in power point format), and no matter the form of slide rule there are no special considerations needed, since the rules for multiplying and dividing do not change despite the style of slide rule. There are even virtual slide rules (see footnotes). Plus there are websites which have slide rule loan programs for a class if a teacher is so interested. Finally one could even download printable scales and have students construct their own slide rule! Imagine making a tool that with the classic 9 scales (C, D, C1, A, B, L, K, S, T) rivals the power of a scientific calculator, is personally hand-crafted, and has such a history. With the basic slide rule, the journey of the mind in acquiring problem solving skills and connecting math and science to the universe can begin.

Web sites

Information: The Oughtred Society : www.oughtred.org
Virtual Slide Rules: Derek's Virtual Slide Rule Gallery :
www.antiquark.com/sliderule/sim
Information, Virtual Slide Rule, Slide Rule power point
presentation on how to use the slide rule, and printable
scales for making a slide rule : www.sliderulemuseum.com
Slide rule plans
Scientific American magazine reference from May 2006
article on slide rules by Cliff Stoll :
www.scientificamerican.com/media/pdf/Slide_rule.pdf

Math & Science Activities : www.cosmicquestthinker.com

Data Analysis Math Tool Alternative Consideration

It is agreed upon today that students need to have connections to the ideas they learn and hands-on activities are the first critical step. The next step then is for them to take their measurements and find a way to connect the numbers to concepts. One of the most important goals of science is to **analyze data** to reach mathematical conclusions and find relations between the variables.

The question then becomes : Is there a different way to examine data? An interesting approach would be one where the students are not only acting as the scientists taking measurements, but also as the mathematician analyzing their measurements. **The answer is the tool, the 'stick' with numbers on it.**

What if, students were to use only low-cost basic tools (rulers, meter sticks, string, thermometers, small masses, marbles, a personally constructed incline made of meter sticks, stopwatches, mass scales, etc) for measurable labs. With basic tools the students take measurements themselves and then with the help of the laws of mathematics and through the use of a 'stick' with numbers on it, the students come to discover and find the relations that they can then read about in their texts?

Even in the case of non-measurable labs where the students are merely supplied, straight-forward data, the students can use the very same mathematical 'stick' and find their relations through some graphing and basic computations.

What 'stick' is this? It is the common slide rule!

Why this tool? The **slide rule** is a tangible and visual bridge connecting numbers to the measured real world. It can be seen as an extension of the use of number lines in their early school journey where they were used for adding and subtracting, only here the slide rule is now used for multiplication and division. The slide rule can also act as a motivation for reasoning and mastery of math.

To use a slide rule, one must first estimate answers mentally, know what and why the measured values used are, sequence the mathematical steps of the problem, and understand their place values through scientific notation of both the variables and the answer.

Learning to read the graduations on the slide rule, (along with learning to use a new tool for calculation) is useful in itself. *Hence, the student becomes the measuring scientist and the computing mathematician simultaneously once again, like those long ago who used such tools.*

The most critical present-day problem, then, is to find such a tool. The references at the end of the article note the International Slide Rule Museum web site, where there is a student-loaner program. For the cost of about $11 per semester, a teacher can be loaned a classroom set of slide rules. There is a power point on how to use a slide rule, along with ideas on its history, and a way to have medals for slide rule competitions as well. Also in the references is a web site, Cosmic Quest Thinker, for suggestions for many science and math activities using slide rules. Each of these has further links to other web sites for virtual slide rules, printable slide rules, publications, even places to assemble one's own classroom set of slide rules and the like.

Data Analysis with a Slide Rule :

In any and all lab situations or even tables of data cases, the students take recorded (or given) data and then merely convert the values into log values of these numbers (read the log value on the L scale from the data value on the D scale on a slide rule). Now they proceed to graph a log-log plot of each of the variables, such as :

- log(displacement) vs. log(time) for constant acceleration cases;
- log(period) vs. log(distance) for pendulums or planets (Kepler's 3rd Law);
- log(Force or Intensity) vs. log(distance) for inverse-square laws (such as gravitational or electrostatic forces or light intensity), et al).

In this new log data set on its graph, draw a best fit line through these points, and then find the slope of the line. The slope taken as the ratio of two simplified whole numbers will show the exponential relation between the variables and the exponents involved.

For example, in the case of constant acceleration (dropped objects or masses on inclines), the ratio of displacement to time will have a ratio of 2 to 1, hence $d \sim t^2$. This means that the graph is of the form $y = x^2$, which is, indeed, a parabola.

Once the variable relation is found, the slide rule can then be used to then check it as well as explore the relation. Continuing the above example, graph now displacement vs. time-squared as well as the log values of each and again for each draw a best fit line and determine slope. The former determines acceleration while the latter should demonstrate a slope of 1.

In the case of inverse-square laws (gravitational, electrostatic), the slope on a slide rule has a ratio of -2 to 1 for Force to Distance. The negative slope is a negative exponent, so it can be seen as $F \sim \frac{1}{d^2}$. This inverse-square law idea applies to light intensity as well.

Even a situation as complex as Kepler's 3rd Law can be examined this way and one finds what Kepler found (using logarithms, no less) that the period-squared is proportional to the distance-cubed for a planet ($P^2 \sim D^3$).

Note that each of these and many more calculations can be done with as simple a tool as a common 9-scale slide rule! This very tool is as powerful as a conventional scientific calculator today.

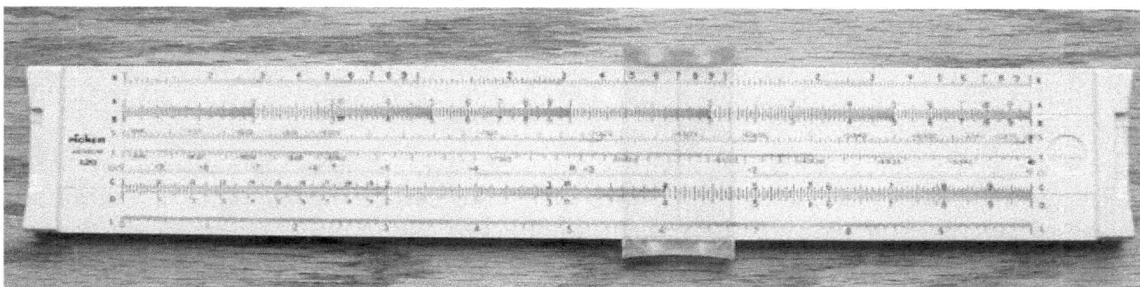

Other mathematical reasons for the slide rule : The slide rule can also be used to illustrate the idea of the *laws of logarithms*, such as the product rule for logs where the log of the product of two values is the sum of the logs of each of the values in question.
(log(A*B) = log(A) + log(B)). Students can compare given values and reach a conclusion here. It is a means to visually conceptualize ideas, such as what happens to variables when one of them changes. **The scales themselves become the variable under consideration.**

Take, for example students given the density of a pure substance. A student places this value on the C scale opposite the left index of the D scale. Now as they read along the C scale, these values represent mass, the numerator of the equation for density ($\rho = \frac{m}{v}$), while the adjacent D scale is the corresponding volume value for that given amount of mass so as to always end up back at the material's density! Other types of data analysis can be done this way.

Average Speed is similarly done. Distance is the C scale while Time is the D scale. For any determined average speed, as one reads along the C scale, one has driven farther, hence more time (D scale) too. Students can be given data here as well to examine, if preferred.

Because *all similar ratios are set up instantaneously*, this same tool can also be used to easily convert fractions into decimals as well as solve any and all proportions even faster than one can on a calculator. Here tables of information can have blanks to be filled in where students can use the slide rule to find the answers. This can be useful for scaling drawings and maps, calculating changes in recipes, determining cost per unit volume or mass, finding unknown sides of similar polygons, and calculating conversions. **The applications are limited to one's imagination and mathematical skill.**

Notice how this idea extends to a simple activity connecting ideas in math and science in the understanding of the value of π. Students can measure circumferences and diameters of common circular objects and find the ratio on the slide rule. It will show π (3.14) if done correctly and since all similar ratios are set up, for a given diameter (or circumference) they can predict the circumference (or diameter). This application applies to any known ratio. The slide rule is naturally poised to deal with any and all ratios simultaneously. This means too that the slide rule is a natural for most formulae students encounter since most are ratios or better seen as proportions, which is the main power of the slide rule.

Other scales of the slide rule allow for explorations and examination of squares and square-roots (A & B scales), cubes and cube-roots (K scale), as well as trigonometric relations of sine, tangent, and cosine (S & T scales). In combination these scales are all that is needed for virtually all formula in science and math through school. With any given formula a student can learn, practice, and master mental math estimation and see the results for herself on the rule with any given calculation. These data values used can all be done as given tables or through measurements depending on the resources and time.

With this approach, using the slide rule, the goal of having the students do the work and discover the outcome is achieved here. The goal of data analysis is achieved. When they do a lab and take the measurements, they now take the data and find the relations using

15

math reasoning when using the slide rule. The students here engage in the art and act of discovery through actually doing the math. The students come to find the various relations either through measured or as given data tables. Along the way, they connect the numbers to real-world phenomena.

Also a startling notion develops – *all values measured can be represented as a number between 1 and 10*, as does the slide rule and this promotes the use of scientific notation. Image their surprise when they realize they are holding infinity in one's hands! The use of the slide rule is just an alternative and a way to inspire a path to mathematical reasoning and understanding. Also consider that the slide rule has a sufficient level of precision with 2 or 3 significant figures, which is all that is needed. The tool helps in reinforcing this idea.

Does this mean the end of the computer or calculator? No. *In fact, the calculator and the computer can act now as a follow-up to check the answers. Instead of being the source of the answers, they are the checking system for the student's work as a follow up.*

What of the use of logarithms and the need to explore them? This can be done in the science or in math class, if the students are at that level for understanding. Otherwise, letting them know that logarithms are a tool to uncover such relations may be sufficient at this time. As noted here, this idea can be extended to any and all other variable relations they encounter in various science classes as well as math classes.

This exploration can be a cool math tool adventure. Students mentally and mathematically examine data themselves to find the answers. They use tools that help them visualize the concepts and make finding answers a personal responsibility and journey. The slide rule promotes math skills acquisition. As an aside, the students can also be introduced to and connected to history through the role of the slide rule. The slide rule was in the hands of numerous scientists, mathematicians, and engineers and used for nearly 350 years (1620-1970). It has a history of being part of the making of the Panama Canal, the Empire State Building, the Golden Gate Bridge, along with development of the steam engine, the discovery of oxygen, and the determination of the density of the Earth. Both Einstein and von Braun used the same 9-scale model themselves – one from the realm of theoretical physics while the other in the practical realm of applied physics to rocket engineering, where he built the Saturn V, the largest human-made device to leave the Earth carrying aloft Apollo astronauts to the Moon, each carrying a Pickett 600 slide rule.

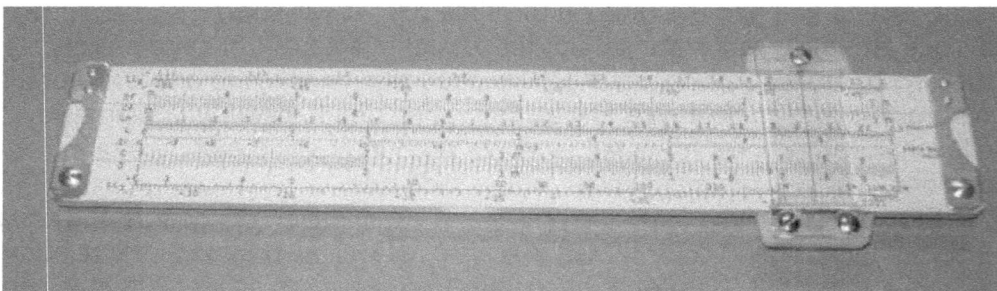

Resources :
Slide Rule Loaner Program, Directions for Slide Rule Use, Make your own slide rule :
http://sliderulemuseum.com
Many Classroom Ideas for Slide Rule use :
www.cosmicquestthinker.com

Ch.III
How to Use a Slide Rule

How does a Slide Rule Work

The Slide Rule is a mathematical tool that enables the user to perform mathematical calculations of a great variety and obtain a reasonable answer. Though it can be used to add and subtract, it is better just to do this oneself. The Slide Rule is best suited for multiplication and division primarily. But with additional Scales the Slide Rule can be used for other things like squaring, taking the square root, cubing, taking the cube root, determining the power or root of a given expression, determining the log of a given number and the inverse of this by finding the number for a known logarithm and in trigonometry can be used to find the values of the sines, cosines, and tangents of given angles along with their inverses where one has the sine of a given angle and needs to find the angle itself. The range of application mostly depends on the user, her math skills, creative but mathematically sound approach to a problem and the number of scales the Slide Rule has to ease the outcome of sought after answer.

The basic parts (and these are noted for the linear model) are the following : The top and bottom strips are called Stators and respectively are referred to as the Top Stator and the Bottom Stator. These pieces are also called Stock in some books. The moving piece between them is the Slide. The Moving Cursor is simple called the Cursor or sometimes called the Indicator or Runner while the cursor line is also known as the Hairline (it was a hair long ago and first suggested by Isaac Newton).

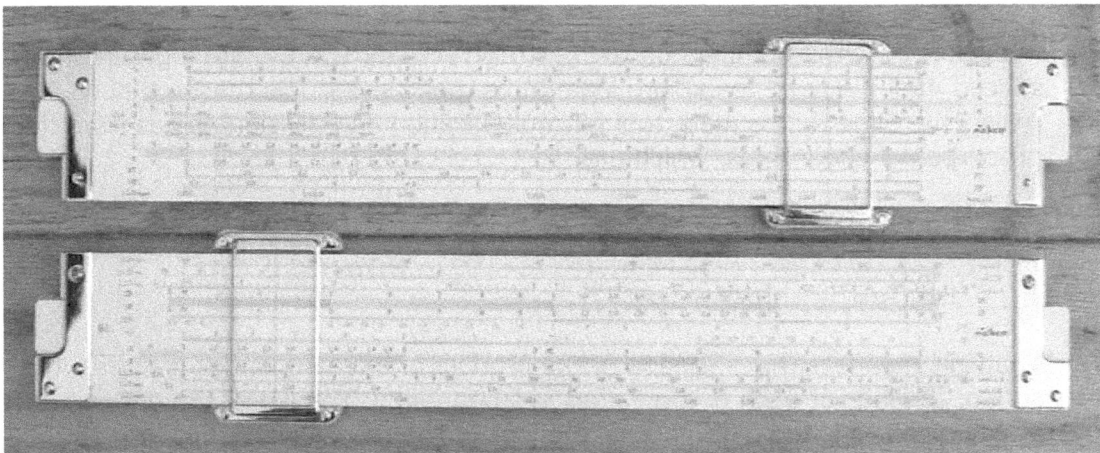

The accuracy of the Slide Rule's answer depends on the Precision of its Scales. Most often the average 10" is effective in its precision of values on its Scales for calculations involving 2 significant digits, but the 3rd sig fig can be estimated as well. The number of gradations on the Scales depends on length, so the longer the Scale the larger the number of gradations, hence a greater level of precision can be found. However, it is not as simple as it might first appear. A 20" slide rule merely has a marginally larger number of gradations hence only a slight increase in estimating the 3rd significant figure. To illustrate : between the 9 and right index 1 of the 10" slide rule there are a total of 10 secondary marks, so values like 9.1, 9.2 are easy enough. Between each of these marks in an ever -narrowing gap (they are spaced logarithmically) on the 10-inch rule there is a mid-point mark, so this would be 9.15, 9.25, et al. It is easy to conclude that one can estimate values to the nearest 0.01, so values like 9.12, 9.37 can be determined where the 3rd digit is estimated while the first two (assuming the slide rule is accurately built and the user is mathematically adept) are reliable. In the case of the 20" rule the same primary and secondary marks exist, but since the scale is longer the tertiary marks now are not 0.5, but instead 0.2. One now has a greater certainty in determining whether the value is indeed 9.13 or 9.14. Notice, however, there is no gain in the number of significant digits though the length of the slide rule has doubled! In general, however, most calculations in the everyday world require only 2 significant digits, so the 10" is sufficient.

The Slide Rule is not a measuring tool, like a ruler, but instead has Scales of Numbers on it that are spaced based on the C & D Scale values which are numbers (such as 1, 1.2, 2, 3.5, et al) placed at a distance from the leftmost number '1' value (aka the Left Index) that corresponds to the logarithm of that given value and multiplied by the size of the scale (in a standard rectilinear rule about the size of a ruler this would be 10" or 25 cm). For example, the number 5 is at a distance from '1' that is log(5)*scale length of the slide rule. This means that all the values are logarithmically spaced from each other on the Scale. What value does having this spacing of values have?

To explore why the numbers on the key base scales (C & D) are logarithmically spaced, first we need to explore logarithms themselves. The reason that it makes the Slide Rule a math tool for multiplication and division comes from the properties of Logarithms, which are :

$$Log(A*B) = Log(A) + Log(B)$$

$$Log(A/B) = Log(A) - Log(B)$$

What this means is this : The log of the product of any two values is merely the sum of the logs of each of the values independently. The log of the division of any two values is the difference of the logs of the respective values as well. What this means is that multiplication can be turned into addition and division can be turned into subtraction. So all we would need then is a table of the log values for any set of values we wish to multiply and we merely add them and find this sum on the table and that would be the product of our values! But the Slide Rule is far easier than this. As noted in the History of the Slide Rule section William Oughtred explored the idea of logarithms and placed the numbers from

1 to 10 at distances from each other that corresponded to the logarithms of the numbers on this line. Knowing the property that the product of any two values is the sum of their logs, all one has to do is combine the distances, hence add them that separate the numbers physically for any given product and the overall total distance will then end up on the number which is the product of the two numbers. Here is a simpler illustration. If I wanted to find the product of 2*3, I merely take the log(2) which equals 0.301 and the log(3) which equals 0.477 and add them together to obtain 0.778 which is the log(6). On the slide rule I merely have to place the starting point, the Index beneath the number in question on another logarithmically-spaced scale – so I place for example the Right Index 1 on the C scale above the 2 on the D scale (go ahead and do this). Now read along the C scale to 3 and look back to the D scale and what do you find? 6, of course. You have traveled the distance of 2 on the D scale and the distance of 3 on the C scale – adding these values together is the same as the product of the values since they are logarithmically spaced so we are to the answer of 2*3 or 6. Read more thoroughly the section on Using the Slide Rule to gain greater insight into using the slide rule for basic and more advanced skills as you learn more and do more with it. Notice we do not need to know logarithms themselves, how to derive them, find tables of them, and with a slide rule in hand, we do not have to construct our own math tool to do this.

Reading the Scale on a Slide Rule :

- All slide rule forms (linear, circular, et al) will have the same basic method in reading the scales
- A Scale is the logarithmic-based spaced numbers or a related line to the base scales (C & D)
- Each mark on a Slide Rule can represent any value as needed through Scientific Notation. So only the values 1 to 10 are needed.
- Watch when doing division or multiplication with scientific notation – see those rules if needed.
- Essentially when values are in scientific notation and are multiplied, then the exponents are added. When values are in scientific notation and they are divided, the divisor (the denominator) is subtracted from the quotient (the numerator).
- All scales are related to the base scales (the C and D scales)
- A base scale has the values from 1 to 10, logarithmically spaced.
- However, each mark has a particular reading based on its place on the particular length slide rule.
- The Numbers on the Scale are the Primary Marks. The next major sets of marks are located between them are called Secondary Marks, and when marked, the marks between these are called Tertiary marks.
- When there are 10 marks between any marks, they are tenths (0.1, 1/10)
- When there are 5 spaces between major marks these are two hundredths (0.02, 2/100)

- When a cursor is between any set of corresponding marks, this value becomes the estimated digit and can be done to the level of 0.1 of the major marks involved at best. If very small distances, then it is best to assume only 0.5 of the value between marks.
- The Primary Mark on all scales is the Index which is the number 1. There is a Left one and a Right one on linear slide rules.
- Linear scales (C & D) have 2 Indexes while circular ones have 1.
- The Indexes on a linear slide rule are referred to as the Right Index and Left Index.
- The majority of slide rules have capacity for 3 significant figures in their calculations. (the range is 2 to 4)
- Best General Rule : Use Scientific Notation in computations
- A, B, C, D, C1, CF, DF, C1F, K, and R scales are all logarithmically spaced values. Some are one time (C & D), others (A & B) are double, some are triple (K), some are reverse (C1 & C1F), some are common square roots (R), some are folded (i.e start at a different point) (CF & DF – starting at π)
- Trigonometric Scales include : S, T, ST
- Log Scales include L, LLN(+ & -) Scales – L is the log value (usually base 10) of a value while the LLN scales are representative of natural log based powers of a C or D scale value to a given exponent (see LLN scales for more information).
- Note, unless noted basic math is done with C & D scales
- Scales written in black are from left to right, while in red are right to left
- All basic scales begin with 1 (except for folded, trig or Log scales).
- Scales are aligned to read across them.
- When a Slide Rule is a duplex (two sided), the cursor can be read for both sides as needed.
- The last digit is estimated in reading a slide rule.
- It is necessary to keep track of the decimal place in calculations.
- Some of the most critical skills in using a Slide Rule are these :
  1. Always mentally keep track of an Estimated Answer (some sort of range for the answer)
  2. Keep track of the Decimal place in using the slide rule, since it has infinity contained in 1 to 10
  3. Mentally visualize the formula under consideration and project this onto the slide rule so as to keep track of which scales to read

Multiplication with a Slide Rule :

- Quick summary of rule (explained step by step below) : Set one index of the C scale to the multiplicand on the D scale. Next, set the cursor of the runner to the multiplier on the C scale. Finally, read the Answer on the D scale under the cursor hairline.
- [Note that the answer can be found on either the C or D scale and it depends where one starts and which index is used – it is up to you, but is determined by whether you go off scale – If the calculation is not present, then use the other Index].

- To multiply with a Slide Rule set the Index of one scale (C) over the first number to be multiplied on the opposite scale (D). [Think of this as X from Q = X*Y]
- Slide the cursor to the second value as it appears on the index scale (C). [This is Y from the equation]
- Read the answer under the cursor's hairline on the opposite scale (D). [This is Q]
- Numerical Example :
- Place the Left Index of the C Scale over '2' on the D scale.
- Now read along the C scale to the value of '3'.
- Opposite 3 on C is 6 on the D scale, which is the answer of 2 x 3. This is because you have moved the scales relative to each other both the distance of log(2) and log(3) when added they are the log(6) distance, hence 6 is the answer.
- What is being done is the addition of the distances of the logs of the values in question
- (Log (X*Y) = Log (X) + Log (Y))
- Though illustrated with C & D scales, note it can be done in reverse (that is alternate C for D and vice versa) and this action can also be done on any paired scale-slide combination, such as A & B or if preferred CF & DF.
- With both multiplication and division (plus other functions) keep track of the decimal point. Using Scientific Notation is the best choice.
- Scientific Notation Method :
-
- In Scientific Notation, when multiplying the exponents are added.
- Special Rule : If the slide projects to the Left when performing multiplication, Add One to the Exponent Value for the correct answer.
- (Note : This is for each calculation, so if done twice, then add two for example)
- (Note : Also very important these rules apply to only the use of these scales in use, C & D, for example and does not apply to reading across to other scales, such as CF & DF for example)
- In Scientific Notation, when dividing the exponents are subtracted. (divisor exponent – dividend exponent).
- One must note, however, whether or not the coefficients when multiplied exceed 10, etc.
- Special Rule :
- Also if the slide in division projects to the right, then the answer from scientific notation exponent needs the subtraction of one from the answer to make it correct. – Note this is for C scale as the Numerator and D scale as Denominator. If the scales used are done in reverse of this, then so too is the rule, hence if it projects to the left in that case then subtract one. (The idea here is the same as it was for multiplication – the number of times the slide extends to the right, for each time – subtract one from the exponent total).

-
- Key to calculation is know your decimal placement from products and ratios and using scientific notation it is always one to the right or left of number on the slide rule!

Division with a Slide Rule :

- Quick summary of rule (explained step by step below) : Set the cursor hairline of the runner to the dividend on the D scale. Then slide the divisor on the C scale under the cursor hairline. Finally, read the Answer on the D scale under one index of the C scale.
- [It is important to recognize that for both multiplication and division the use of the scales can be reversed. Here in step one, C is the numerator while D is the denominator. These roles can be reversed].
- To divide with a Slide Rule set the divisor value on one scale (C) over the dividend value on the opposite scale (D).
- Read the answer above the active (D) scale index (1) on the opposite scale (C).
- If the answer falls outside the range of the scale (such as when multiplying or dividing numbers from opposite sides of the scale) then the other index needs to be used in a linear slide rule.
- (Note : This does not occur with a circular slide rule).
- Numerical Example :
- Place '6' on the C scale over '2' on the D scale.
- First note that we are creating a ratio of 6 over 2.
- Next realize that we need to use the Index. In the case of division, always use the Index on the Scale that is the Denominator – here that is the D scale for the example, but it can be the other way if you wish.
- Now notice that the Right Index of D is not opposite any value, so we need to examine the Left Index of D, which is opposite the value of '3' on the C scale, the answer to 6/2.
- What if we wanted 2 divided by 6?
- Here the denominator is the C scale, so again find the useful Index, which is the Right one in this case.
- Since 2 < 6, we expect our answer to be less than 1.
- The right Index of C is opposite 0.333 on the D scale.
- Notice that 3rd digit – it comes from the fact that each division between 3.3 and 3.4 is 0.02 and the cursor is in the mid-point between 3.32 and 3.34. Finally following the rules of scientific notation and decimal placement, since the slide went to the left and each of the values has an exponent of 0, so 0-0 is 0 and then we subtract one from that and have -1 for our exponent, so our answer of 0.333.
- What is being done is the subtraction of distances of the logs of the two values in question
- (Log (X/Y) = Log (X) – Log (Y))
- The best way to consider this operation is think of it as a proportion : the Answer is over 1 while along the scales the Divisor is over the Dividend.

-
- The Proportion concept is useful for calculations involving conversions and manipulation of 3-variable functions.
- With all slide rules where one scale can move while the other is immobile all answers are present in both multiplication and division as the scales are read. This makes it a parallel calculator – all relations are instantly set up and visual.
- Though the Scientific Notation method works best for values, be wary and if the slide projects to the right and you are reading the Right Index, then in using the Scientific Notation system for the values, subtract one from the answer in all cases. If it projects to the right, then it is alright for division, yet remember that for Multiplication you add one to the total exponent value!
-
- Alternative Multiplication and Division Method :
- Characteristic Method (which is akin to the Scientific Notation method, only here the characteristic is the exponent) :
- For any given set of values, write the characteristic (the portion in front of the decimal when written as the log of the value in question (recall the notation : **characteristic.mantissa** – Note only the characteristic is needed, we are not looking up the mantissa, which, by the way, can be found on the L scale of your slide rule if needed) –
- Be sure to be wary of positive and negative values in this case!
- Sum up these characteristics.
- Now perform the Multiplication or Division with the Slide Rule as you normally would. That is to say, each number is merely a value from 1 to 10 only. Note that the answer may end up on the range of values before 1 to 10 (i.e. be 1/10 as large) or on the range after 1 to 10 and be in the range of 10 to 100.
- For multiplication, each time the slide , with the C Scale, extends past the left index of the D scale, add one (+1) to the Sum of Characteristics.
- For Division, subtract one (-1) from the Sum of Characteristics.
- The revised total is now the characteristic (i.e. the power of 10) for the answer to use with the value showing on your slide rule.
- Return to our 2 x 3 and 6/2 examples for a moment. Each has a Characteristic of 0. In the case of multiplication, it went to the right, so the sum is again 0, while in division the sum is 0, but we subtract one yielding an exponent of -1 from these rules here.

Combined Operations with a Slide Rule :

- Combined operations are multiple operations in one problem
- In this list there are many examples and notated use of the Slide Rule when it comes to fractions, ratios, proportions, and applications of these ideas.

-
- Let's say we have the situation for continued products where
  Q = A x B x C x ---
- Here set the hairline of the cursor indicator at A on the D scale.
- Next move index of C scale under the hairline.
- Next move hairline over B on the C scale.
- Now move the index of C scale under the hairline.
- Next move the hairline over C on the C scale
- Now move the index of C scale under the hairline.
- Continue moving hairline and the index alternately until all the numbers have been set in this case and come to the answer.
- Second scenario :
- If the problem reads (N x M) / R then
- Place N (C scale) over R (D scale)
- Slide the cursor along the D scale to M (on D)
- Find the Answer on the C Scale
- Repeat this process as needed for more than this set up
- Be sure to keep track of Decimal Point as noted above in the Multiplication and Division Rules.
-
- In essence, combination problems can be seen as proportions which are extensions of ratios, which slide rules are good at.
- For example, any two values over each other is a Ratio or Fraction and once set all similar ratios are automatically established instantaneously. As well opposite the Index of the Divisor is the decimal equivalent of the ratio as well!
- Convert the decimal to a fraction -
- <u>What of converting from a Decimal to a Fraction?</u>
- Place the decimal value on the C Scale over the Right Index of the D Scale and then search along the C and D Scales for a ratio of numbers to represent them.
- Keep in mind this decimal is the percentage value, but you need to multiply in your head by 100 to see it as a percentage.
- For example, 3 on the C Scale over 8 on the D Scale has the Right Index of D under 375 on the C Scale.
- Since 3 < 8, the value is less than one and should be read as 0.375
- What of the Scientific Notation Rules. Each initial value has the exponent 0 and in division, 0 − 0 is 0. But the slide is projecting to the right so we subtract 1 from the result and have -1 for an answer. This means to move the decimal one to the left, and the answer is read 0.375
- As a percentage, multiply by 100 and the answer is 37.5%
- <u>One of the largest uses of fractions is for Sales!</u>
- Place the Price of the Item on the C Scale over the Right Index of the D Scale.
- Read backwards along the D Scale to 9, 8, etc. Each of these is read as N x 10% (such as 90%, 80% and so on). Above it is the Price at that percentage!

-
- Of course, keep in mind, this is not the percentage off, but what you are paying. To see how much is saved, just use the Left Index of D Scale under the Price instead and read to the right instead of left.
- For example, 25% is found at 2.5 and the value above it is 25% of the price. (Of course what you pay is found at 7.5 or 75% instead).
- Sales tax and total cost can be found in a similar manner :
- Place the Cost of the item over the left index of the adjacent scale 9 say C over D as we have been doing)
- Read along the D scale to the sales tax expressed as a decimal value (this should not be too far, 4% is 0.04, 6% is 0.06, et al)
- The value above on the C Scale is the total cost including sales tax!
-
- Other Fractions or Ratios in Everyday Life :
-
- Mpg = miles per gallon
- Take the ratio of miles driven over the number of gallons used sometime to determine just what gas mileage you are getting!
- Mph = miles per hour
- What was your average rate of travel for a trip?
- Take the ratio of the Miles driven to the amount of time (in hours) to find the average speed of the trip!
- What if you are some fraction of the distance there and have determined the average speed, then
- The question becomes how much longer 'to grandma's house?'
- Take the remaining distance and divide by the average speed to estimate the number of hours to complete the journey.
- Further, from the mpg calculation you could take the remaining distance divided by the mpg and find the number of gallons needed for the trip in order to decide whether to fill up again or not!

- The fraction as a Slope in Algebra
- A ratio might not be just two numbers, but instead represents the ratio of two differences of numbers :

- $m = \dfrac{\Delta Y}{\Delta X}$

- This is the slope of a linear line, one of the largest topics in algebra.
- The slope is the rate of Rise Over Run. The larger the value, the steeper the line.
- The sign of the slope determines whether the line is moving up or down when examining it from left to right.
- You can consider slope as a 3-variable formula (see below).

- The Fraction as a Rate :
- In many practical applications of the fraction it is seen as a **Rate**. The numerator is one value (distance, gallons, cubic feet of gas, amount of

- growth, temperature change, etc) while the denominator is some other value (typically time).
- The rate could be determined or it may be known in a given problem. If to be determined, the amounts of the other two variables are known
- If the rate is known, then clearly we are missing either the amount in question flowing or the amount of time needed to do this.
- A very good example of a Rate is in Cost per Unit Ounce (Volume), etc. This is a valuable tool for the slide rule in comparing items when shopping for their comparative costs to find the better deal.
- The flow rate is how water and natural gas consumption is measured and billed. For water and natural gas are Ccf (100s of cubic feet).
- In order to estimate the Cost, one has to merely read the meter at the beginning and the ending times, subtract to arrive at an amount used and multiply this by the cost per unit. This is true for all of the meters : Water, Electrical, and Gas.
- Rates and Ratios are very common in Science :
- In Physics there are many ratios, such as Density (amount of substance per unit volume), Pressure (force per unit area), Power (Joules per second), etc.
- Also in Physics, rates can be speed (rate of change of distance with respect to time) or acceleration (rate of change of velocity with respect to time) or from Newton's 2nd Law acceleration is the ratio of Net Force to the mass of the object undergoing the net force.
- Chemistry has many ratios such as Molarity (the number of moles of solute to liters of solution), Molality (the number of moles of solute to kilograms of solvent), Percent by Volume [Mass] (the volume [mass] of Solute to the volume [mass] of solution), the Law of Definite Proportions where the Percent by Mass (is Mass of the Element divided by Mass of the Compound), as well as the Law of Multiple Proportions (these could be looked at in the Proportion section obviously), and so on.
- In Chemistry rates are seen in things such as rate of reaction (is the negative of the rate of change of reactant to change of time) plus many more.

- Conversions :

- In Conversions there are two basic methods that can be used with the Slide Rule :
- First, treat the two items as a Fraction that is to be multiplied by the Number in question.
- Typically the ratio of the items is taken where the units one is 'in' are in the denominator, while the units one wants to convert into are in the numerator.

- $$\frac{\text{Units to Convert Into}}{\text{Units to Convert From}} \quad \text{or} \quad \frac{X}{Y}$$

- This results in this situation :

-
- Beginning Value in Initial Units* $\frac{\text{Desired Units}}{\text{Initial Units}}$ = Final Value

- $N * \frac{X}{Y} = M$

- The best way to handle this on a slide rule is to :
- Place N on the C Scale over Y on the D Scale on the slide rule
- Read along the D Scale to X, the Desired units and read the answer, M on the C Scale.
- Looking at the prior discussion, it is easy to see that the formula presented can be read as :

- $\frac{M}{N} = \frac{X}{Y}$

- All one has to do is take the ratio of Convert Into Units on the C Scale over the Convert From Units on the D Scale
- Now read along the D Scale to the Beginning Value (N) and find the answer above it on the C Scale (M)
- Use the following Table and look elsewhere for everyday conversions. It is best to keep a small list and memory of these as needed :

- **1 inch = 2.54 cm**
- **12 inches = 1 foot**
- **3 ft = 1 yard**
- **1 mi = 5,280 ft**
- **1 minute = 60 seconds**
- **365 days = 1 year (rounded)**
- **1 solar day = 24 hours**
- **1 cup = 8 fluid ounces**
- **1 gallon = 4 quarts**
- **1 pound = 16 ounces**
- **1 kilogram = 2.2 pounds**
- **1 ounce = 28.3 grams**
- **1 liter = 1.06 quarts (rounded)**

- Plus look up whatever you may need !

- For each of the above and any others your find a need for, simply place one value over the other as described above it for use.
- <u>Conversions are needed in many applications :</u>
- One basic unit to another (inches into feet),
- Changing one unit type into another (English to Metric),
- Currency Conversions, and others.
- <u>Computing Costs</u>

-
- The reason for conversion depends on the needed outcome. For example feet into yards is commonly used when computing the area of a room in square-yards for carpeting by dividing by 9.
- In the case of painting, the Slide Rule readily calculates the area of a wall with Length times Width, but what of the number of gallons of paint needed?
- Take the total area to be covered and use the Slide Rule to divide by 300 if the walls are unpainted or rough – if smooth and already painted divide by 350 – to determine the number of gallons of paint needed!

- In Math, angles are often converted between Radians and Degrees :

- $180° = \pi$ radians

- In the Sciences, there are numerous conversions not only of the aforementioned units, but also of mixed units :

- Such as km/hr to m/s is very common.
- For that calculation the ratio for it is 3.6/1. Check it yourself!
- In Chemistry there are conversion commonly found in the amount of substances and how it is expressed :
- For example take the number of grams of a substance and divide by its gram molecular weight to determine the number of moles present.
- In Science and Math there are many other conversions depending on the situation at hand.
- In Math in the realm of trigonometry since the C and D Scales are the values for the Sines and Tangents as read from the S and T Scales (keeping in mind where the decimal falls), this makes it easy to multiply or divide by the sine or tangent of a value when and where needed.
- Slide the sine value for a given angle read on the C Scale from the S Scale over a given value on the D Scale.
- Read in one way the sine is in the numerator and in the opposite direction it is in the denominator.
- What if you want to find the Sine or Tangent value and are given the sides of a triangle?
- Obviously this is again a ratio : For example :

- $$\sin\Theta = \frac{\text{Length of Side Opposite}}{\text{Length of Hypotenuse}}$$

- Also other angular measures are available and can be used for determining distance or size, such as in the Radian measures :

- $$\textbf{Radian measure} = \frac{\text{Arc Length of Circle Portion}}{\text{Length of Radius}}$$

- If our protractor measures 1/10th of a radian, then the apparent size of the object in question is 1/10th the measure of the distance from us!

- Another interesting one in Math is the conversion from one base to another in terms of logarithms.
- The standard slide rule has base 10 logs, but let's say you want a number in another base, say 2 or the natural log base e (2.71828*)?
- For any positive numbers, N, A, and B with A ≠ 1 and B ≠ 1,

- $$\log_A N = \frac{\log_B N}{\log_B A}$$

- From the question the natural log of any value is the ratio of the log base 10 of the number divided by the log base 10 of the natural log value.

- To illustrate, here is a similar example question :
- What is the log base 2 of 6, for example. $\log_2 (6)$
- Look up reading from the d Scale to the L Scale both the log of 2 and the log of 6. (0.301 and .778 respectively)
- Now divide on Scale C and Scale D 778 by 301 - We find it to be 258
- Where does the decimal go?
- Since 6 is much greater than 2, it will have a characteristic (here 2 and the rest is the mantissa 0.58) so the answer is 2.58
- So $2^{2.58}$ = 6 (try it and see, rounded off of course)
- The idea of the ratio, fraction goes on to any and all applications.

- **The Proportion :**

- As noted in the prelude, the proportion is two ratios set equal to each other.

- In the conversions section above, it is easy to see its value in use.
- The basic proportion can be expressed as :

- $$\frac{M}{N} = \frac{X}{Y}$$

- All one has to do is take the ratio of known values - M on the C Scale over N on the D Scale

- Now read along the D Scale to the other known value (Y) and find above it on the C Scale the answer (X)
- It is easy to see that all one needs to know is any 3 of the variables and the 4th is the one to find.
- By using the scaled as fractions it is easy to place one value on one scale and the other on the opposing scale.
- Try this for yourself to find that solving a proportion on a Slide Rule is indeed much faster than one can solve one on a Calculator!

-
- Also here there are no complex rules, like the calculator, such as cross-products –
- In the case of the Slide Rule the natural form is maintained which s the equivalence of two ratios.
- This is an invaluable tool in Algebra for proportions as well as in geometry for any and all similar figures to determine unknown sides!
- Proportions can be used to determine height or distances - In this case we are using similar triangles.
- For example, hold up a ruler at arm's length to measure the apparent height of a distant object, say a picture on the wall.
- If you read the apparent height, measure your arm's length, (this is the first ratio)
- now measure the distance to the wall,
- Set the first ratio of your measures to the ratio of the unknown height on the wall to the distance to the wall.
- Then the height of the picture can readily be determined.
- Though simple, this technique is used in surveying regularly and is used in the wilderness for distances across rivers and gorges before the advent of electronic equipment.
- The easiest way to envision it is when you are outside and you as well as a tree or a flag pole casts a shadow on a sunny day.
- The ratio of your shadow length to your height is equal to the shadow length of the flag pole to its actual height.

- $$\frac{\textbf{Length of your shadow}}{\textbf{Your Actual Height}} = \frac{\textbf{Length of flagpole shadow}}{\textbf{Actual Height of flagpole}}$$

- A more thought out activity involves determining the size of the Moon :

- Use a meter stick and place a small card vertically with a hole punched in it at a distance
- Look along the stick through the hole so that when viewing the full Moon it fully fills the diameter of the hole (i.e. move the card until you have proper alignment)
- The ratio of the diameter of the hole to the distance that the hole is from your eye equals the actual diameter of the Moon to the Moon's distance from you. (Here, we assume we know the distance to the Moon). Hence the diameter of the Moon can be determined!

- Proportions can also be used in changing the scale of a recipe :

- $$\frac{\textbf{Recipe Requirement for Material}}{\textbf{Recipe Number of Servings}} = \frac{\textbf{The amount of Material Needed}}{\textbf{Number of Servings}}$$

- Here the ratio of how much is needed to the number of servings is set equal to the amount of unknown material and the number of desired servings. The amount needed is readily found.

- **In Math there are many applications :**

- The very nature of the proportion stems directly from geometry and the relations found in similar figures.
- These can be the aforementioned triangles but also includes any similar polygons, such as squares, rectangles, and the like where one is known, the other is partially known and there is a missing side.
- Still other examples are considered :
- *What if one wants to find the circumference of a given diameter of a circle   (or vice versa)?*
- Simply put π on the C Scale of the slide rule over the Index on the D Scale. – Note with CF & DF scales this is already done since these scales are set at π as their beginning point over the C & D Indexes! See CF & DF scale use
- Now read along the known scale. The C Scale is the Circumference, while the D Scale is the diameter.
- *What about the diameter if the area of the circle is known?*
- Take the Area on the A Scale over p on the B Scale.
- The diameter-squared is found on the A Scale opposite the B Index.
- The diameter is then found below reading from the A Scale to the D Scale to read the square root of the value on A.

- Other interesting things can be done with gauge marks (special marks on slide rules for conversions, multiplications, et al like π) as well as personally derived values :

- On some Slide Rules there is a mark, c, on the C & D Scales (1.273) which is $\frac{4}{\pi}$ and comes from $A = \pi * \frac{d^2}{4}$
- Place this C mark or value of the C Scale over the index of the D Scale
- Slide the cursor over the size of the diameter of a considered circle on the D Scale.
- To find the Area of a circle read the answer on the B Scale!
- What if you want to do this by knowing the radius (of course we could simply multiply by 2 to use the former method, but give your mind a chance to explore the Slide Rule !)
- Look up the square root of π by first finding it on the B Scale and noting its value on the C Scale.
- Slide the square root of π value on the C Scale to the Index of the D Scale
- Now read along the D Scale to any desired radius value for a circle.

- The answer in this case is read on the B Scale for the Area of the Circle in question!

-
- What about the Volume of a Sphere?
- Now place the value 1.61(2) on the D Scale over the index of the C Scale.
- Read along the D Scale, for the radius of a given sphere you are considering.
- The Volume of this sphere is found on the K Scale!
- This comes from $(\frac{4*n}{3})^{1/3}$ from the formula $\mathbf{V} = \frac{4*n*r^3}{3}$
- There are many other problems in Math and Algebra in the area of problem solving:
- For example, if a given material costs so much per ounce, pound, ton, how much will another desired amount cost ?

- $$\frac{\textbf{Cost}}{\textbf{Unit Amount}} = \frac{\textbf{How much does it Cost?}}{\textbf{Amount Wanted}}$$

- There are numerous ratios of values that can be found or derived from many other references that can be used in proportions to find the answer to many a question one might encounter in a math and or science text:

- $$\frac{\textbf{Diameter of Circle}}{\textbf{Side of Inscribed Square}} = \frac{\textbf{99}}{\textbf{70}} = \frac{\textbf{Diagonal of Square}}{\textbf{Side of Square}} = \textbf{(2)}^{1/2}$$

- Here is a general value for the pressure one feels with depth :

- $$\frac{\textbf{Pounds per Square Inch}}{\textbf{Feet of Water}} = \frac{\textbf{26}}{\textbf{60}}$$

- Even more complex proportions can be solved :

- A classic algebra question might read :
- *If it takes 4 people 7 days to accomplish a job, how much time is needed (assuming the same work rate) for 6 people to do this task?*

- $$\frac{\frac{\text{Initial Workers}}{\text{1 Task}}}{\text{Amount of Time Initally}} = \frac{\frac{\text{Workers in case 2}}{\text{1 Task}}}{\text{Amount of Time needed}}$$

- $$\frac{Wi}{\frac{1}{Ti}} = \frac{Wf}{\frac{1}{T2}}$$

- $$\frac{4}{\frac{1}{7}} = \frac{6}{\frac{1}{X}}$$

- You could go through and first simplify it and then take the ratio of the numbers on one side once the variable is isolated, but the slide rule allows for this to be solved as is!?

- Take 4 on the D Scale and slide 7 on the C1 Scale over it. Start with the cursor here.
- Read along the C1 Scale to 6 and look below on the D Scale to find the answer 4.66 days.

- **In Physics, for example, say you have a balance beam.**

- *If on one side of the balance you have 26 g a distance of 32 cm from the center,*
- *how much mass must be placed on the opposite side at a distance of 20 cm from the center in the opposite direction so that it balances?*

- cw is clockwise, ccw is counter-clockwise

- $$\frac{\text{Mass cw}}{\text{Mass ccw}} = \frac{\text{Distance ccw}}{\text{Distance cw}}$$

- $$\frac{26\ g}{X\ g} = \frac{20\ cm}{32\ cm}$$

- X = 41.6 g

- This idea can be applied to problems in chemistry too.

- Take for example, conservation of mass in a problem where one has to determine the mass of a reactant product in a total mass size where one is only given a small sample for testing.

- $$\frac{\text{Mass of reactant in sample}}{\text{Mass of Sample}} = \frac{\text{Mass of reactant in total mass}}{\text{Total Mass}}$$

- This list applies to any and all sciences and is limited only by the imagination in the questions being asked.

Using the CF & DF Scales (πC & πD) :

- The CF and DF Scales are called Folded Scales since they do not start at the Index, but instead are at a chosen point, namely here being π.
- Why π?
- Simple – there are many calculations that involve π that extend from Circles, such as the circumference of a circle :

-
- Take any value on the C or D scale and now look to the cursor value on the corresponding CF or DF scale. It is p times greater. This is the formula for the Circumference of a Circle : $C = \pi * d$, where d, the diameter of the circle is being read from the C or D scale and its circumference is then found on the CF or DF scale.
- For example – put the cursor of your slide rule on 3 on the D scale and find the DF scale. We are assuming we have a circle with a diameter of 3 units, so the question is : what is that circle's circumference? On the DF scale we read the answer of 9.42 units
- Note that the reverse is true as well. If we know the circumference of a circle, we can read it on the CF or DF scale and find its diameter quickly on the C or D scale.
- Another important role that a folded scale has is this : Since it starts at another point on the line, it is easier to do multiplication and division without having to change up Indexes as often.
- For example let's try 3 x 6
- We might first be inclined to slide the Left C Index over 3 on the D scale before realizing we should have first chosen the right C Index.
- But no bother –
- Read the value of 6 on the CF scale and find opposite it on the DF scale our intended answer, 18!
- What of the decimal point rules here, however?
- Going back to our original rules,
- First estimate the answer : Anything past 3.x will result in a value greater than 10, since by 3x4 we are at 12 already.
- Next, if we were to only use the C and D scales, then we would have had to use the Right Index instead of the Left one, hence we would have added one to our exponent total, or 1 in this case (since the values started with exponents of 0).
- Another way to think of this problem is this : Since we have gone past the end of the scale we are on and moved on to the next one in sequence, it is 10x larger, so instead of being 1 to 10, it is now 10 to 100, so the value of 1.8 is really 18 in this case.
- Note : You can do multiplication and division with the CF and DF scales and the same rules apply for them as they did for the C & D scales presented previously. Realize that the index is still 1.

Using the C1 Scale ($\frac{1}{x}$) :

- This is the Inverse Scale. – See an example of use in combination problems – this is a common use of the C1 scale.
- It is basically C scale in reverse.
- It is written right to left (regular are left to right)

-
- The inverse of any number can be found by aligning a cursor over a given value of C and on C1 is the inverse. (Be sure to keep track of the decimal!)
- For example if reading '5' on the C scale, on the C1 it must be 0.2
- To multiply with the C1 Scale set one of the numbers on C1 over the other number on D.
- Read the Answer on the D scale under the Index on C1 scale.
- Any problem with a fractional component can be examined more easily with C1. – It can be a fraction or be the denominator of a given expression.
- Especially useful in combination problems. – see prior for example in the combination section under proportions.
- Use the rules for multiplication and division and read the proper index for a solution
- One of the other important and common uses of the C1 Scale is to be the scale when reading for a value from the tangent scale angles above 45° (45° to about 80°)

Using the A & B Scales (Squares X^2 & Square Roots $\sqrt{X}$) :

- The A & B Scales are double scales and are the squares of C & D Scales –
- That is to say it is double logarithmic (1 to 10 and 10 to 100)
- Conversely C & D are the square root values of A & B
-
- Rules for Square Roots of Numbers :
- Note Left side is Left Index to Middle Index, and Right side is Middle Index to the Right Index
-
- Odd Number Digit Rule :
- For Whole Numbers > 1
- If the number of digits left of the decimal point in the value being considered for square rooting is odd, Read the value from Left-Hand side of A
- For Numbers 0 < x < 1
- If the number of zeroes to the right of the decimal is odd then Read the value from the Left-Hand side of A
-
- Even Number Digit Rule :
- For Whole Numbers > 1
- If the number of digits left of the decimal point in the value being considered for square rooting is even, Read the value from Right-Hand side of A
- For Numbers 0 < x < 1
- If the number of zeroes to the right of the decimal is even then Read the value from the Right-Hand side of A – Note this includes no zero at all (just .X)

- To summarize the rule (and make reading a R scale easier) :
- The Left hand side of the A & B scales is for Odd Number of Digits or the Odd Number of Zeroes in a Number (this corresponds to the R1 scale)
- The Right hand side of the A & B scales is for Even Number of Digits or the Even Number of Zeroes in a Number (this corresponds to the R2 scale)

For 0 < X < 1									
No. of zeroes in between X and the decimal	0	1	2	3	4	5	6	7	Continue the pattern of even then odd values
Which Side (L or R)?	R	L	R	L	R	L	R	L	Continue pattern R,L,etc
√ Answer and the no. of zeroes between ans. And decimal point	0	0	1	1	2	2	3	3	Continue Pattern 4,4,5,5,etc

Number to have a square root taken	Answer on slide rule
0.2	0.447
0.02	0.141
0.002	0.0447
0.0002	0.014

For X > 1									
No. of whole digits in value X	1	2	3	4	5	6	7	8	Continue the pattern of even then odd values
Which Side (L or R)?	L	R	L	R	L	R	L	R	Continue pattern R,L,etc
No. of Digits in the answer √	1	1	2	2	3	3	4	4	Continue Pattern 5,5,etc

Number to have a square root taken	Answer on slide rule
2	1.41
20	4.47
200	14.1
2,000	44.7
20,000	141.

Using the K Scale (Cubes X³ & Cube Roots ∛X) :

- The K Scale is a triple scale and is the cube of the values on C & D
- That is its range is 1 to 10, 10 to 100, and 100 to 1000.
- Rules for Taking Cubes :
- For the Rules consider the K scale divided into 3 sections from one index to the next.
- The sections are Left, Middle, and Right
- For All Values under consideration for cube root extraction :
- Divide the Number into groups of 3 (starting at the decimal point) and go left or right of the decimal point as needed in creating these groups
- Look at the Left-most group with non-zero digits in it
- If the group has 1 digit then use the Left portion of the K scale (for example 1-10)
- If the group has 2 digits then use the Middle portion of the K scale (for example 10-100)
- If the group has 3 digits then use the Right portion of the K scale (for example 100-1000)
- To continue further, continue counting in 3's.
- An easy way to find where a value falls if <1 :
- Let right-most 1 be 1 and go backwards as powers of 10 for any given decimal value :
- Example 0.00X is in the thousands place
- The right-most '1' is 1000/1000,
- The next '1' left of it is 100/1000,
- The next '1' left of it is 10/1000
- And the left-most is 1/1000
- If there are more restart at right-most and continue (Note the scale wraps around then)
- The number of groups left of the decimal point determines where the <u>decimal point</u> in the <u>answer</u> falls :
- If there are two groups (of 3) left of the decimal point (complete or not !) (means values 1,000 – 999,000) then there are 2 figures left of the decimal point in the answer.
- Values : X,XX0 to XXX,000
- If there is one group of 3 (values 1-999) then the answer has one value left of the decimal point
- Values : X.XX to XXX.
- If there are 3 groups to the right of the decimal point (where one or more falls in the tenths-hundredths-thousandths columns) then the answer has 3 figures right of the decimal point starting just past the decimal point
- Values : 0.XXX to 0.00X XX-,---
- Values : 0.000 XXX to 0.000 00X XX0 ---
- If the number has four groups to the right of the decimal point where the first group is all zeroes and at least one value is in the ten-thousandths,

hundred-thousandths, or millionths place, then the answer will have 4 figures to the right of the decimal point where the first is a zero (0).

Decimal value for X 0 < x < 1		Number of Zeroes in value		
Which portion of K scale to read?	R, C, L	R ,C ,L	R ,C ,L	R ,C ,L
Q number of zeroes in value X	0,1,2	3,4,5	6,7,8	9,10,11
Number of zeroes in answer to $\sqrt{}$	0	1	2	3

whole values for X X > 1		Number of Digits in value		
Which portion of K scale to read?	R, C, L	R, C, L	R, C, L	R, C ,L
Q number of digits in value X	1,2,3	4,5,6	7,8,9	10,11,12
Number of whole digits in answer to $\sqrt{}$	1	2	3	4

Using the L Scale (Log Base 10 (N)) :

- The L Scale is the Log value of the C & D scales
- It is effectively used for determining powers and roots for a wide range of values. Note that the powers do not have to be whole numbers as well as roots can be any fractional value one is interested in determining
- The L Scale provides the log of values 1 to 10 with no changes
- If the Number is <1 (includes <0) or >10, the log value has both the Characteristic (the value before the decimal point) and the Mantissa (the value after the decimal point found on the scale)
- Note that the mantissa will be the same for a given value independent of the decimal point.

-
- For example the log (2) mantissa is 301, as is for log (20), log (200), etc – the only difference is the Characteristic, so log(2) = 0.301, log(20)=1.301, log(200) = 2.301, et al
- When the number is <1, can use scientific notation to determine the value to add to the log of the number in question. Log(0.2) = Log(2 x 10^{-1}) = 0.301-1 = -0.699

-
- **Procedure below for logs in general :**
- First Look up the Log value for any given number treating it in Scientific Notation format so that it is >1 and <10
- Locate the characteristic of the scientific notation value on the C or D Scale and look below to the L Scale
- This L value is the Mantissa
- Add the Mantissa to the Exponent of the Scientific Notation exponent
- This Sum is the Final Answer
- The L Scale is useful for X^N and $X^{1/N}$
- To solve – Look up X on C and find its log value on L
- Then Multiply or Divide as needed the N or 1/N value involved in the problem using the rules for multiplication or division. Be sure to watch the decimal point.
- This new value, Q, is computed
- Now Search for the Q Mantissa on the L scale
- Note that if there is a Characteristic, it becomes the power of 10 for decimal placement of the answer!
- Alternative to finding a log for values between 0 and 1
- If a desired value for log(X) has 0<X<1 as a decimal in the tenths place, the C1 scale can be used. Reading 2 on C1 the log value is the log of (0.2) for example.

Using the S Scale (sin(θ) & sin⁻¹(θ)) :

- The Sine (S) Scale has values representing the angles from approximately $5.5°$ to $90°$
- When the cursor is on any value of the S scale (the left number printed in black typically) its sine value can be found under the cursor on the C or D Scale
- Note : Since Sine functions range from 0 to 1, all values read from C or D are decimal and begin as 0.XXX (for most models)
- Note : Some Box-style Post Slide Rules have a cursor on the backside where the S scale often is and with a value on it the sine & tangent (T scale) of the angle is found on the B scale!
- For angles $0°$ to $5.5°$ there is often an ST scale and represents both Sine and Tangent functions as these have approximately the same values for small angles such as these (also the decimal is 0.0XXX)

-
- Many Sine Scales have a second set of numbers written in red to the right of the values on the scale –
- These are the complimentary angles hence they represent the Cosine of those angles listed in red
- Recall sin (Θ) = cos $(90°-\Theta)$
- To multiply by the Sine of a given angle simply place the angle over the appropriate D scale index
- Next read along D to the value to be multiplied by
- Find the Answer on the C Scale above that point on the D scale
- Remember to watch the decimal points since the C value is a decimal value
- To divide by the sine of an angle simply place the angle on S scale acting as the dividend over the divisor on the D scale
- Find the Answer at the C scale corresponding D scale Index
- If the square to the sine of an angle is needed since the sine value is on C, the square of the sine value is found on A.
- To find the log of an angle for a trigonometric function such as sine, look up the sine of the angle on C scale, then read on the corresponding L scale for the log of the sine of this angle.

Using the T Scale (tan(θ) & tan⁻¹(θ)) :

- The Tangent (T) Scale has values printed in black from 5.5° to 45° when read left to right
- The values (typically red) read right to left are 45° to 84.5°
- Like the directions for Sine (S) Scale,
- To read the tangent (Θ)
- place the cursor on the desired angle
- If the angle is from 5.5° to 45°, the Tangent to this angle is found on the C or D scale and is read as a decimal 0.XXX
- Recall that tan(45°) = 1.00
- In fact this value lines up with the Right Index of C & D
- If the angle is 45° to 84.5° read the T Scale from Right to left to find the value and place the cursor there
- Answer to the Tangent of this angle is found on the C1 Scale
- The values on the C1 scale are read as whole numbers as the line appears
- (Note there are slide rules that have two T lines, hence are read from left to right on the C scale and one must keep in mind that the C values start at 1 (tan 45°)
- Like the S scale since the tangent (like the sine value) value is on the C scale the rules for multiplying and dividing are the same as noted in the S Scale section
- Note since tan(Θ) = 1/cot(Θ) one can compute these as needed

Using the ST Scales :

- If your Slide Rule has a ST Scale it can be used for small angles for both the sine and tangent function as these are very close to each other in value between 0.6° and 5.5°
- Much like S Scale and T Scale readings the angle is on the ST scale and the reading comes from the C Scale.
- Important Reading Scale Note : Here however the readings range from 0.01 and 0.1, so the reading has 10^{-2} as its exponent.
- What if there is no ST Scale :
- In approximations :
- We can use : $\sin(x) = \tan(x) = \frac{x}{\frac{180}{\pi}} = \frac{x}{57.3}$
- Note that this approximation also is the radian value for that angle!
- Here x is the small angle and using the C and D scales then provides a reasonable value for the sine or tangent of the small angle (within the same range).

Using the Log (LLn Scales) :

- The LLn Scales are used to raise a number in question to a power or find a root of that number.
- Each line is a power of e (2.718...), some models it is the power of 10, of the next line in the list (LL1, LL2, et al).
- That is to say each line is e to some power (from -10X (LLO3) to +10X (LL3))
- The Table of Names and Powers for LL scales are below showing both the power and the range of values to be found on those scales. :
- Looking at the tables, it is clear that the spacing needs to be examined carefully when reading the scales. Be sure to look at the two primary values that your values is between (as an example) and then look at the number of secondary divisions to determine the appropriate value.
- These scales are indispensible when looking for any given power or root as needed (not necessarily whole number ones either).
- Also since each is 10x the line before it, for example LL3 is 10x LL2, and LL2 is 10x LL1, and so on, hence LL3 is 100x LL1. This means that for any power of 'e' which is used for the scale when reading from the C or D scale each is 10x the others.
- For example e^2 is found by looking at the D line for 2 and reading from the e^{1-10x} line, which is LL3 and is approximately 7.4
- Better still, what if you wanted the inverse of e^2 or 7.4? Simple read its value on LL03 – Be careful in reading the scales, these are in reverse first off and have decimal values – so read it from right to left
- Here the inverse of 7.4 is between 0.1 and 0.2 and reads 0.135 – try this yourself and practice reading the scale properly.

-
- What about $e^{0.2}$, that is found on the LL2 scale, (yields a value nearly 1.222)
- and $e^{0.02}$ is on the same cursor line with the cursor the whole time on 2 on D.
- This is one of the great values of the LLn scales. Any value from 1.001 to nearly 100,000 can be used from LL0 to LL3 and its inverse is present as well if needed. That would be 10^{-5} to 0.999 found on LL00 to LL03. It is typically written in this range though, due to the effectiveness of the slide rule : having a range of values (up to 80 inches in length on these LLN scales) of running from 0.00005 to 20,000.
- Looking at a value on say LL0 scale and reading the value below on LL1 scale it becomes that number raised to the power of e 10x more. So an increase in number is 10x the prior line.
- The LL0 value if read with the LL2 scale will then be that value raised to the power of 100.
- Going from a higher LL scale number to a lower one means that it s $1/10^{th}$ power per each jump.
- To raise to a negative (−) exponential power : This can be done directly or find the positive value then its inverse on the corresponding negative exponent line.
- If raising negative exponent to positive exponent values :
- If one wants to find 10x the value in power simply read from one line to the next, such as going from LL1 to LL2 in essence multiplies the power by 10 if needed.
- Going in the reverse direction divides the power by 10, of course.
- For example read a value on −LL1 scale and find that value to the 10th power on −LL2 scale.
-
- Note : the −LL Scales are normally written in red. And the values increase from right to left (as do all other inverse scales)
- Recognize that −LLX scales are the inverse of LLX scales correspondingly. So the inverse of the values found on LL1 are found on LL01 also known as −LL1.
-
- **Using the LLn Scales in general :**
-
- To find X^N :
- Look up the value X on the LL Scale.
- Slide the Index of the C (1) scale over it.
- Move the cursor along the C Scale to the power (N) and
- Read below the answer on the needed LLX scale
- For example, 2 is squared, 3 is cubed, 5 is the 5th power − but you can do any power in between too − such as 4.7^{th} power for whatever reason is needed :
- (be sure to estimate since it may have gone from one of the LLX Scales to the next one in line)

-
  - **One could find roots as well :**
  - For a given value, X find this on the needed LLX scale
  - Place the needed root over it (N) on the C scale
  - Note : Recognizing the this is read as 1/N mentally
  - Now read in conjunction with the C scale Index the value on the appropriate LLX scale the root value
  - Essentially taking the root is the reverse of the process of determining a power – much like doing multiplication and division with the slide rule in terms of directions.
  - Note : this process works for –LL Scales as well.
  - LL Scales are not good for numbers very close to 1, such as 1.001 or 0.999.
  - There is an approximation for this value for small values of 'n' with : $(1 + n)^p = 1 + d^p$

Name	Power	Range	Name	Power	Range
(+)LL3	$e^{+1.0x}$	e to 22k	(-)LL03	$e^{-1.0x}$	1/e to 1/22k
(+)LL2	$e^{+0.1x}$	1.1 to e	(-)LL02	$e^{-0.1x}$	0.91 to 1/e
(+)LL1	$e^{+0.001x}$	1.01 to 1.1	(-)LL01	$e^{-0.001x}$	0.990 to 0.91
(+)LL0	$e^{+0.0001x}$	1.001 to 1.01	(-)LL00	$e^{-0.0001x}$	0.999 to 0.990

Note : in table, the + and – denotation is used together, such as –LLN and +LLN, on some slide rules, while on others it is LLN and LL0N denotation depicted

Tricks & Tips in Slide Rule Use :

- Check alignment of scales and adjust as needed
- Always be sure of the level of precision of the scales for proper reading. Most 10" Slide Rules are quite good to 2 significant digits with even a 3rd estimated significant digit being possible.
- In terms of all calculations undertaken, Be sure to first estimate the answers ahead of time mentally
- Often it is a good idea to convert most values to scientific notion both for ease of calculation and determination of decimal placement.
-
- **Summary of Scientific Notation Rules :**
- **The Rules for Decimal Placement :**
- 1) Always first estimate the Answer.
- 2) Convert all values into Scientific Notation.
- 3) For Any Multiplied Values, Add up the Exponents and For Any Divided Values, Subtract the Exponents.
- 4) If for any single Multiplication operation the Slide moves to the Left, then add +1 to the exponent total (Why? Because in essence, we have gone off this scale and are adding two values that extend beyond the scale in front of us to a next one in line, which is 10x the line before it)
- 5) If for any single Division operation and the Slide moves to the Right, then -1 from the exponent total (Why? Because in essence, we have gone off this scale and are subtracting two values that place the answer on the scale to the left of the one in front of us, which is 1/10th the line before it)
- 6) Mentally treat any two values independently (just as whole numbers now) but keep track of what the exponent will be through the rules.
- 7) That is to say answer the question as to where the Slide has moved and factor in that addition or subtraction as needed for each operation independently.
- 8) Take this final answer and use the exponent figure arrived at to determine the decimal placement!

- **Alternative Decimal Place Method :**
- **Summary of Decimal Rules using the Counting Digits Method found in some slide rule books :**
- The key to the digits method is to tally the number of digits in a number.
- The basic rule for this is this :
- For numbers X > 1 the number of digits is simply the number of digits in the number.
- For example : 7 has 1 digit, 70 has 2 digits, and 70,000 has 5 digits, etc.
- What if the number is greater than 1 but has decimals, though?
- For example the number is 23.45, how many digits does it have?
- It has only 2.
- In the case of X > 1, all decimal values are overlooked in the digit count.
- So the next question then is, what of values 0 < X < 1 then, what is their digit count?

-
- In the Digit Count Method all values 0.1 to 0.9 expressed as nothing more than 1/10ths values have 0 (zero) digits.
- So 0.6 has 0 digits, for example.
- With each decimal place it is like a reverse number line :
- 0.01 has -1 number of digits,
- 0.001 has -2 number of digits,
- 0.0001 has -3 digits,
- And so on...
- To help remember this the value of digits is essentially the negative number of zeroes past the decimal point.
- Summary of these ideas on Counting :
- For numbers greater than 1, count all the numbers up to the decimal and treat these values as positive. For numbers less than 1, count the number of places up to the first nonzero number and view these as negative values. Sum up these numbers in multiplication and when the slide extends to the right, subtract one from the sum.
-
- With the number of digits in all the values you have in your problem now look to the following rules when it comes to multiplication and division :
-
- Multiplication with C & D Scales :
- 1) If the slide projects to the right of the stock during multiplication, the digit count for the product is one less than the sum of the digit counts for both the values in your calculation (called the multiplicand and the multiplier).
- 2) If the slide projects to the left of the stock, the digit count for the product is equal to the sum of the digit counts for the multiplicand and the multiplier.
- Division with the C & D Scales :
- 1) If the slide projects to the right of the slide rule stock during a division, the digit count for the quotient is one more than the digit count for the dividend (the numerator) minus the digit count for the divisor (the denominator).
- 2) If the slide projects to the left of the stock, the digit count for the quotient is equal to the digit count for the dividend minus the digit count for the divisor.
- Multiplication with the C1 & D Scales :
- 1) If the slide projects to the right of the stock during multiplication, the digit count for the product is equal to the sum of the digit counts for the multiplicand and the multiplier.
- 2) If the slide projects to the left of the stock, the digit count for the product is one less than the sum of the digit counts for the multiplicand and multiplier.
- Division with the C1 & D Scales :
- 1) If the slide projects to the right of the stock during division, the digit count for the quotient is equal to the digit count for the dividend minus the digit count for the divisor.
- 2) If the slide projects to the left of the stock during division, the digit count for the quotient is one more than the digit count for the dividend minus the digit count for the divisor.

-
- Other Ideas :
-
- Examine the problem carefully, working from the inside out, and use the best scale for that calculation
- Recognize that values can be found easily with given scales – squaring, square rooting, cubing, cube roots, sine values, etc. When moving the cursor its alignment may find it at a place to make use of these as needed.
- Proportions (which also conversions and 3-variable functions can be treated as) are very straightforward and easy to set up and solve on a slide rule.
- Note that most things are Ratios or Proportions and the Slide Rule is the best tool for these calculations since it sets up all similar ratios instantly with any setting!
- Create a needed scale! Take the activity where we used the C & D scales with the C1 scale. If there is no C1 scale, simply invert the slide in the slide rule and use the reversed C scale as if it were a C1 scale now.

- **How to Add Numbers on a Slide Rule :**
- Suppose we have two numbers X & Y so that we want their sum (X + Y) and yet use a slide rule!
- Let's first rearrange this expression : $X*(1 + \frac{Y}{X})$
- With a slide rule simply set the Y value on C over the X value on D.
-
- If Y > X read answer above the D index from the C scale as whole number with decimal and add one to it. Note that the decimal placement rules also must apply however.
- OR
- If Y< X read answer at D index on the C scale as decimal value and add one so that it is 1.***. Note that the decimal placement rules must also apply however.
-
- Now Take the result and find it on D and place this value under the left C index.
- Now, Regardless of Y > X or Y < X, now read along the C scale to the X value and find the answer for the sum of X + Y on the D scale below! Be certain to have an estimate of your answer and employ the proper decimal placement rules for reading the slide rule.

- **Other Things to find on the Slide Rule :**

- There are many ratios that can give good estimates in situations :

- Set up these ratios and find the known value and opposite it is the sought after answer -

$$\frac{\text{Circumference of Circle}}{\text{Diameter of Circle}} = \frac{355}{113} \qquad \frac{\text{Feet}}{\text{Meters}} = \frac{82}{25} \qquad \frac{\text{Atmospheres}}{\text{Feet of Water}} = \frac{23}{780}$$

$$\frac{\text{U.S Gallon}}{\text{Cubic inches}} = \frac{1}{231} \qquad \frac{\text{Feet of Water}}{\text{Pounds per sq inch}} = \frac{60}{26} \qquad \frac{\text{Side of a Square}}{\text{Diagonal of a Square}} = \frac{70}{99}$$

$$\frac{\text{US gallons}}{\text{Liters}} = \frac{14}{53} \qquad \frac{\text{US Gallons}}{\text{Imperial gallons}} = \frac{6}{5} \qquad \frac{\text{Inches of Mercury}}{\text{Feet of Water}} = \frac{15}{17}$$

$$\frac{\text{Yards per Minute}}{\text{Miles per Hour}} = \frac{88}{3} \qquad \frac{\text{Pounds per sq yard}}{\text{Kgs per sq meter}} = \frac{46}{25} \qquad \frac{\text{Weight of fresh water}}{\text{Weight of sea water}} = \frac{38}{39}$$

$$\frac{\text{US Gallons of Water}}{\text{Weight in pounds}} = \frac{3}{25} \qquad \frac{\text{Ounces}}{\text{Grams}} = \frac{6}{170} \qquad \frac{\text{Diameter of Circle}}{\text{Side of equal square}} = \frac{79}{70}$$

$$\frac{\text{Inches}}{\text{Centimeters}} = \frac{26}{66} \qquad \frac{\text{Pounds}}{\text{Kilograms}} = \frac{75}{34} \qquad \frac{\text{Area of Circle}}{\text{Area of Inscribed Square}} = \frac{322}{205}$$

Ch.IV
Gauge Marks and Scales of the Slide Rule

The Scale

     The primary use of the slide rule is seen in our activities for multiplication and division. This was done by using non-linearly divided scales that are divided instead in a logarithmic manner (that is to say the numbers were not evenly spaced like on a ruler but here by logarithmic distances).

     Why use the logarithmic spacing of the numbers ? The most basic linear form utilizes two scales where the numbers are logarithmically spaced. Due to the mathematical properties of logarithms, the spacing of the numbers on these lines allows for easy and rapid multiplication and division with the same set of rules despite the type of slide rule used. (This idea is explored in the history section in more detail discussing the discovery and importance of logarithms to history which the slide rule can be seen as the physical visual manifestation of this math form).

     Many slide rules have more than 2 scales, and these can be used for many other mathematical operations as we have seen, such as : squaring, square roots, cubing cube roots, raise to various powers, taking a various root of a number, common and natural logs, and trigonometric values such as sine, cosine, and tangent.

     Standard scales spaced in a logarithmic fashion, like C and D are said to be **Single Scales**. That is, there is one run from 1 to 10 in the distance for the slide rule, such as 25 cm (i.e. 10 in). If there are two scales running from 1 to 10 in the same distance, it is said to be a **Double Scale**, such as for A or B scales. Of course if there are 3 runs of 1 to 10, which is the case for the K scale, this is a **Triple Scale**.

     Another type of scale is the **Folded Scale**. Instead of starting at the Index point of 1, it begins at or near another point on the logarithmic line. The usual choice is at pi (π). The choice of pi was two-fold. One it allowed for using pi as an Index, so that calculations involving circles and cylinders could easily be done and two, by cutting the original C or D scale at a different point, made alignment more convenient for times when multiplication involved numbers from each end of the regular fundamental scale.

     The C and D scale are the primary or fundamental scales of the Slide Rule and all others are based on them. For example, we see from the activities, the tangent (as well as the sine) scale is read in relation to the C or D scale. This idea is further illustrated in the table which lists in alphabetical order the scales, their mathematical relationship to C or D and a brief description of that scale.

Table of Scales and Uses

SCALE	Mathematical Relation to C or D	DESCRIPTION and Other Notes
A	X^2	Square Values of Fundamental D scale Double scale, Opposite to B scale
A1	$1/x^2$	Reciprocal of A scale, Reciprocal of square of D scale
B	X^2	Square Values of Fundamental D scale Double scale, Opposite to A scale
C	X	Fundamental (Single) Scale On Slide opposite to D scale
CF	πx	Folded Fundamental Scale, starts at π Opposite to **DF scale**
CI	$1/x$	Reciprocal of Fundamental Scale On Slide – Basically a Reverse Scale
CIF	$1/\pi x$	Reciprocal of CF scale Reciprocal of Folded Fundamental Scale C
D	X	Fundamental (Single) Scale On the Stock opposite to C scale
E	e^x	Log-log scale – see LL3
K	X^3	Cube Values of the Fundamental D scale Triple Scale
L	$Log_{10}x$	Mantissa of the common logarithm of Fundamental D scale value
LL	$Ln(x)$	Mantissa of the natural logarithm of the Fundamental D scale value
LL0	$e^{0.001x}$	Scale yields 'e' raised to 0.0001*x power, where x is read from the fundamental scale. Positive Log-log scales are used to raise a number to some exponent or find roots >1 LL scales are to enable one to raise values to a power and take roots very readily
LL1	$e^{0.01x}$	Scale yields 'e' raised to 0.01*x power, where x is read from the fundamental scale.
LL2	$e^{0.1x}$	Scale yields 'e' raised to 0.1*x power, where x is read from the fundamental scale.
LL3	e^x	Scale yields 'e' raised to x power, where x is read from the fundamental scale.
LL00	$e^{-0.001x}$	Scale yields 'e' raised to (-0.001*x) power, where x is read from the fundamental scale. Negative power means $1/e^{-0.001x}$ Negative exponent log-log scales are used to raise numbers to a power or find roots <1
LL01	$e^{-0.01x}$	Scale yields 'e' raised to (-0.01*x) power, where x is

		read from the fundamental scale. Negative power means $1/e^{-0.01x}$
LL02	$e^{-0.1x}$	Scale yields 'e' raised to (-0.1*x) power, where x is read from the fundamental scale. Negative power means $1/e^{-0.1x}$
LL03	e^{-x}	Scale yields 'e' raised to (-x) power, where x is read from the fundamental scale. Negative power means $1/e^{-x}$
P	$(1-(0.1x)^2)^{1/2}$	Pythagorean Scale. Cosine of $\sin^{-1}$ D scale
R1	$\sqrt{\ }$	Square root of Scale D value Scale twice the length of D scale R1 runs 1 to 3.2, R2 runs 3 to 10 Aka: W1, W2 and Sq1, Sq2
S	$\sin^{-1}x$	**Scale D** value is the sine of the angle in degrees read on the **S scale** Runs 5.7° to 90°
ST	Sinx, Tanx	Sine & Tangent of small angles 0.58° to 5.73° Same and S&T scale Used since sine and tangent are similar at these angles
T	$\tan^{-1}x$	**Scale D or C and CI** value is the tangent of the angle in degrees read on the **T scale** Runs 5.7° to 84.3° D scale read for angles 5.7° to 45° increasing CI scale read for angles 45° to 84.3° (printed in red and in reverse order)

The table above is by no means all of the scales, nor does it trace the differences in names for the same scales as they changed through time. It is a representation of the most common scales and represents a list that most slide rules commonly have a subset of.

As noted earlier, one could simply use a 9 scale form (such as in the activities) and be able to compute the overwhelming majority of problem types that even a regular scientific calculator is capable of today.

This does not mean that the other scales were merely for show, though there was probably some pride and showing off in the office if one had a more advanced model I imagine. The other scales were applicable to various disciplines, such as chemistry, physics, electrical engineering, and the like.

This idea is particularly true when one considers not just the right scale combination form, but also relevant and useful gauge marks that may be on the scales. Some are listed and described below in the accompanying table. Like the scale table, this is not complete and represents a cross-sectional view of them.

Table of Gauge Marks on the Slide Rule

Besides the numbers on a slide rule perhaps you have noticed some out of place marks or letters? These are gauge marks. These marks are at specific places corresponding with their value and have a given prescribed mathematical value to the user of the slide rule.

The table below lists some of the more common ones. This table is very far from complete but illustrates some of the possibilities. Note that not all makers use the same letters nor do they have all of the same marks on their various models. Some were probably put on there since those models may have been marketed to a particular set of professions where that mark would have value and use.

Look at some of the examples such as symbols for the weight of copper conductors, watts in one horsepower which clearly had applications in the electrical industry. Other constants as gauge marks can include the acceleration due to gravity for calculations in physics.

More common ones include the conversions of radians to degrees and vice versa as angular measures in calculations were commonplace.

Probably as no surprise the most common gauge point is π, since it can be used for all circular calculations. Using π we realize the value in this. Knowing where it is at allows quick calculations of either multiplication or division as needed.

Gauge Marks	Meaning	Value
C	Square root of $4/\pi$ Circle area calculations	1.128
G	Acceleration due to Gravity (metric)	9.8 m/s^2
G	Acceleration due to Gravity (English)	32.2 ft/s^2
L	Natural log of 10 Convert $\log_{10}$ & $\log_e$	2.3026
R, p, or r	$180/\pi$ Convert radians & Degrees	57.3
p′	Minutes in a radian	3438
Q	Radians in one Degree	0.01745
	$\pi/4$	0.7854
π	Pi – ratio of circle Circumference to Diameter	3.1416
W	Weight of copper	111000
-	Watts in one Horsepower (hp)	745.47 (746)

There are also common conversion marks and value marks with no special symbols on various slide rules, such as on the Fowler's Circular Calculators. Many of them have value marks for square-root of 2, square-root of 3, along with pi, and conversions for inches to centimeters, kilograms to pounds, square centimeters to square inches, and the like.

Ch V
The Story of Energy Essay

This following essay is a lengthy story of Energy and most of the central ideas concerning it. This is being done in order to introduce all of the Activities in the book on Electricity and Magnetism. This is because electricity is not only an energy form that is central to everyday life both in electrical power sent to our homes and businesses but also in battery form in many devices we not only use but rely upon. Not only is electricity so universal but then there is the problem of how it is generated. You can read this through first then move on to the Activities with these central ideas in mind, or come back to it as needed when doing various Activities while engaging in the measuring and calculating for yourself when ideas and/or questions arise. Enjoy.

The definition of Energy

All of us have heard of the concept of Energy, but often when asked we have a hard time defining or describing it. We know we need it, and there seems to be a wide variety of types. It is in the paper and/or news most days. One of the key things is this – energy is essential. In matter of fact, Energy is probably the most central of all concepts to all sciences, since nothing is done without energy basically. The best road to Energy begins with Work. **Work**, by definition, is a Force acting through a Displacement.

$W = F*d*cos\Theta$
   If $\Theta = 0°$,(i.e. they are in the same direction) then $W = F*d$

In all cases for Work being done, there must be at least two things :
1) There is a Force acting on an object and
2) there is a displacement or movement of that object (AND noting that the force is acting in the direction of the displacement as well).

This definition of work leads to the famous question posed in Physics. If you exert a force, say 10 N on a book and the force is parallel to the table. You push the book a distance of 10 cm. Have you done any work and how much? From our formula, it would be :

$W = F*d$ ($\Theta = 0°$ here, so $cos\Theta = 0$)
$W = 10$ N $*0.10$ m $= 1$ N$*$m $= 1$ J

This introduces us to the unit of Work, namely the newton*meter, commonly referred to as the **Joule**, named after James Joule who was a physicist who worked in heat-energy relations. This, however, is not the 'trick' question in physics. If you hold the book in the palm of your hand and exert 10 N straight up to hold it in place (i.e. balanced forces) and now step 10 cm forward, have you done any work and how much?

Before you say that these are the same numbers, hence the same answer, read it carefully. The force is at an angle to the displacement. In fact it is at 90°. The cos(90°) = 0, hence W=0! We have done no work according to the rules here.

Does this mean that there is Work or Not only. Obviously not, since there are a wide range of angles between 0° and 90°. Each of these has some amount of work according to the formula. In many introductory physics classes the forces are considered to be in the direction of the displacement, hence the need for angles is not considered.

Returning for the moment to the **Joule** – how much is 1 Joule? Simply put, we would have to move a 1 N object 1 m and therefore we have done 1 J of work. (For English units : 1 N is approximately 0.22 lbs and 1m is about 3 ft. So lift a ¼ lb. apple one yard and you have exerted the equivalent of 1 J of work). In essence, a Joule is a small amount.

How does this relate to Energy? So in this case here, looking at our first 'work' calculation, we pushed a book horizontally. If after that 10cm push we stopped pushing, what would happen? According to Newton's Laws, it should travel now at a constant speed unless affected by an external force. How fast depends on its mass. Push a large massive book, it would move much more slowly than a lightweight book. The question is, can we find a relation for how fast the book should be traveling?

Let's use the Equations :

$W = F * \Delta d$
$W = m * a * \Delta d$

Our goal is a relation having only mass (m) and speed (v)
We need two formula to substitute for 'a' and 'd'.
To further simplify, let $d_i = 0$ and $t_i = 0$

$a = \dfrac{\Delta v}{\Delta t}$

$d = v_{av} * \Delta t$

$v_{av} = \dfrac{v_f + v_i}{2}$

Substitute for 'a' and for 'd' and also put in for 'v_{av}' in 'd'. With some algebraic manipulation, you end up with :

$$\mathbf{W = \tfrac{1}{2} * m * (v_f{}^2 - v_i{}^2)}$$

We define '½*m*v² ' as Kinetic Energy (abbreviated KE)

W = KE$_f$ - KE$_i$

From this we have the general equation :

K.E. = ½*m*v²

Before pressing on, how fast is our book moving? Recall that we had 1 J of work done to move it. Let the mass be approximately 1 kg (you'll see why in the potential energy discussion below). The speed then turns out to be about 1.4 m/s (try it for yourself on the slide rule).

Kinetic Energy is the energy due to motion of an object. Looking at the Work relation to it, we have what is now called the
Work-Energy Relation.

Work is now defined as the change in energy of a system. If the change of energy is positive, then the system has taken in energy. If the value is negative, it has given off energy.

W = Δ(KE)

Can we do work to something, yet not change its overall motion once we are done doing the work? For example, we take our book in the aforementioned example and now lift it to a book shelf that is 2 m above the floor. Recall we needed to exert 10 N of force to hold the book. From Newton's 2nd Law we know that the force due to gravity is what we call weight and is expressed as :

F_w = m*g = 10 N = m*9.8m/s/s
F_w ~ 1 kg

First note, this is where the approximate mass for the book in the prior example came from. Notice here we have a force and a displacement, the needed elements for work to be done. Second, note we are moving the force in the direction of displacement (unlike our prior 'trick' question).

W = F*d = 10N*2m = 20 J

We also know that Work is the change of energy, and it is positive, so we have added energy to the book, yet it is not moving. Where, then, is the energy? Consider, if the book were to fall off the shelf, would its speed change? Certainly, due to the force of gravity, its velocity changes at the rate of the acceleration due to gravity. If we were to calculate the amount of energy it has as it reached the height from which we lifted (the 2 m) it would have 20 J and we could determine its speed at that point. The energy we put in has returned in the form of motion of the falling object.

This stored energy is what is called **Potential Energy**. Potential Energy, by definition, is the energy associated with an object because of the position, shape, or condition of the object. In this case, it is specifically **Gravitational Potential Energy**. Gravitational Potential Energy has to be measured from a **Reference Point** (a base or starting point). We measured ours from where we lifted the book 'from' and 'to'. Most often in books, the reference point is the ground, where the potential energy is defined as zero. (Note also unless specified, if there is no descriptor in front of potential energy, it usually implies gravitational).

P.E. = m*g*h

W = Δ(PE) = m*g*Δh

Since there is an adjective describing the noun, there must be others. There is Elastic Potential Energy as found in stretched rubber bands, compressed and stretched springs, and any other matter that can snap back when tension forces are applied to it. By definition this is the energy available to use when a deformed elastic object returns to its original configuration. There is also Chemical Potential Energy that is similar yet deals with the arrangement of atoms and molecules. There are other common forms of energy too such as Heat, Sound, and Light all of which can be expressed and analyzed mathematically as Energy.

These first two energies (KE and PE) and their subsets constitute what are called Mechanical Energy (there are other potential energy forms too – these are the most commonly referred to here).
Mechanical Energy is the sum of the Kinetic Energy and all of the Potential Energy of a material.

M.E. = K.E. + Σ P.E.

Mechanical Energy can be conserved. That is to say, what there was initially will be there at the end. Recall the discussion of lifting the book to the shelf and it falling. The amount of energy put in was all that was available when it fell to its starting height.

Initial Mechanical Energy = Final Mechanical Energy
$ME_i = ME_f$

In many cases, there is often only one potential energy form under consideration, such as gravitational potential energy, so we can write :

$KE_i + PE_i = KE_f + PE_f$

There appears to be many variables in this equation. But we need to choose the best places for analysis to simplify the situation. At the highest point for the book on the shelf, it is not moving, so there is 0 KE here. When it has reached the bottom of the fall, the PE is zero and all of it is now KE only!

Of course the question then is how much of it is KE and how much is PE say halfway down? One-half the energy is KE and one-half is PE here. The further the book falls, the more of the energy is manifested as KE, since the speed is changing. The closer to the top of its path, the more PE it has, hence less KE.

This idea is best illustrated by the classic roller coaster. A motor driven system pulls the cars to the top of the hill, hence they have their maximum gravitational potential energy. As they travel down the hill, their potential energy decreases, and their kinetic energy increases. The wheels have low rolling coefficient of friction values so that all the successive hills are easily managed by the car train. Notice that the later hills are shorter than the first big hill. This is not only a good heart-racing effect, but the energy gained there is used to let the cars go through the whole of the track with little to no input of additional energy. In fact, there is enough energy so that the momentum of the system is still there at the end and a brake is applied to bring the cars to a safe stop.

Despite a description that seems very easy, this is the basis of all science and its analysis of energy flow through a system, whether it is in chemistry, biology, physics, astronomy, geology, or meteorology. In the case of energy production, we will dam up a river to create a lake behind the dam, divert some of the water down a tube to hit and spin turbine blades that are attached to a generator to produce electrical power. So the gravitational potential energy becomes the kinetic energy of the water, which in turn becomes the kinetic energy of the turbine, then this is the kinetic energy of the rotor in the generator to produce electricity, which generates the electric field in the wires which becomes the kinetic energy of the electrons in the wires to our homes and businesses.

This all leads to the **Law of Conservation of Energy**. *It states that energy cannot be created nor destroyed, it can only be transformed or transferred from one form to another. The total amount of energy in a system does not change. Like other conservation laws this applies to a closed system.*

The energy of a dropped object turns into heat and sound once it strikes the floor, like if our book we placed on the shelf should fall. Some of the energy can also deform the book too. But most importantly, all of the gravitational potential energy turned initially into kinetic energy which then turned into these other forms of energy. The total at the beginning is the same as the total amount of energy at the end of this situation.

Since the energy cannot be destroyed it is there in the universe. Mostly as heat, which is another important energy topic.

Heat

Why is the study of heat so vital? The obvious answer is nature and weather itself! Land and Water heat at different rates and the angle of the Earth's tilt with respect to the Sun plus its rotation creates a very complex system that results in different rates of heating along with the fact that the incoming radiation is partially reflected into space before even entering the atmosphere (about 30%). Different materials absorb at different rates and the energy can be redistributed, reabsorbed and retransmitted in the infrared (concerns about global warming due to materials even in the air such as CO_2, et al). The amount of water vapor, dust, along with the angle the sun strikes the surface, and the texture, and composition of the terrain plus the wind speed will affect the surface temperature. All of these affect the atmosphere's energy content, hence affecting pressure and pressure differences is the driver of the winds along with other factors too. This drives the weather!

Closer to home, heat and its transfer is critical in immobile structures (homes, buildings, et al) and even in mobile structures (cars, trucks, engines, et al). In buildings it is the retention of heat, particularly at cold times that is of importance. Yet when it is warm outside, the retention of cold air and the cooling of the air is critical. The measurement, dissipation, and control of the heat is very important. Even with a computer and its cooling fans and other appliances, such as refrigerators, and those that use heat like toasters, ovens, and coffee pots, heat once again becomes a factor of importance. The understanding, control, and use of heat is essential to many mechanical and biological items, aspects of life, and life itself.

The study of heat is a branch of physics called thermodynamics . What of the definition of heat and the units of heat? In physics, heat is considered the energy transferred from one body to another due to thermal contact due to differences in temperature. Naturally heat transfers from a body of warmer temperature to one of cooler temperature. When bodies are at the same temperature they are said to be in thermal equilibrium. When they are not the process of heat (energy) transfer can take place. The examination of heat looks not only at appliances and buildings, but also weather, the interior of the Earth, and other bodies in the universe such as the Sun and other stars. Heat is a measure of the quantity of energy of a system, hence there is no 'cold' per se, merely the absence of heat as compared to another body. Also the Heat or Energy can only transfer via one of 3 known means : Convection, Conduction, and Radiation.

Conduction is transfer of heat/energy from particle to particle in a material or between materials in contact with each other. Often conduction occurs in solids. Metals are good conductors while wood typically is not. Convection is the transfer of heat/energy due to the movement of the particles of the substance itself carrying heated particles, such as currents in fluid materials (which includes liquids and gases). Radiation is energy/heat transferred by electromagnetic waves.

The units of heat are : Joule, calorie, kilocalorie. In fact, the factor noted before is the conversion for calories and joules : 1 calorie is 4.186 joules. Note that these are not the food calories we consume, each of these is 1000 of the base unit calories, hence the food ones are often written with an upper case 'C' as Calories. We could then convert our food calories to kilojoules by multiplying by the conversion factor. In many European countries, cans and boxes of food do not note the Calories, but instead the kilojoules instead.

Heat is something which may be transferred from one body to another according to the 2nd law of thermodynamics – essentially it moves from a higher temperature to a region of lower temperature naturally (i.e. the noted natural process and via one of the aforementioned means or if it does go from a place of lower energy to one of higher energy, it requires the input of work).

Consider the question, 'how long will it take my cup of coffee to cool off?'. Quickly any of us would answer that it depends on many factors, or variables speaking scientifically. The amount of coffee, the initial temperature of the coffee, and the type of cup it is in and whether this cup is covered or not would be the first quick responses. With thought, we might add that the ambient temperature of the surroundings plays into it, as well as whether we have added anything to the coffee, such as cream.

The question of cooling is not just the thought of the coffee aficionado, but one of science since the amount of energy a material has and the transfer of energy (heat) concerns many of us whether it involves our homes, businesses, and other buildings, not to mention heat transfer in motors and other mechanical devices, furnaces, and understanding nature, such as questions of how long does it take a lake to freeze or thaw? The natural rate of heat flow is a science unto itself (thermodynamics) and involves topics in the areas of physics, chemistry, astronomy, geology, biology, as well as engineering. The science of thermodynamics is critical as it gives us the conservation of energy (energy cannot be created nor destroyed, only transferred from and/or transformed into other forms). It stipulates the maximum effective work and the limits to efficiency of a system, which coincidentally comes from the temperature difference of two items in a system (Maximum Work (W) $\sim \Delta T = T_H - T_C$)

The question of the Rate of Cooling was one considered by one of the most famous scientists known, Sir Isaac Newton who is more well known for his work in physics, astronomy, and mathematics with his laws of motion, law of gravity, describing the use of a prism to separate light, creating the first reflecting telescope and calculus, and here even his thoughts on 'cooling'. It resulted in what is referred to as "Newton's Law of Cooling". Newton made the observation in his statement that the rate of the heat flow from one body to another (always hot to cold naturally and requiring work to go the other way – this is one of the Laws of Thermodynamics) is proportional to the difference in temperature between the 'hot' and 'cold' objects and time. What this statement implies and seems to defy common sense is that very hot objects cool more rapidly than warmer objects in the same setting. What seems to be a simple expression in words, is actually complex mathematically as expressed below.

$$\frac{\Delta T}{t} = k\Delta T$$

At first glance this idea seems potentially linear, but it is not. It turns out to be a differential equation with a natural logarithmic 'decay' (meaning negative) rate.

Since we are dealing with heat, hence energy flow, we need to consider what helps us transfer this energy and are there any consequences of the transfer and/or transformation of a given energy form.

Simple Machines & Efficiency

Are there ways to change the equation for conservation of energy ? What of **Simple Machines**, such as the lever, the ramp, the pulley, the screw, and the wheel & axle? All simple machines can either redirect force and/or multiply the amount of force (at the expense of displacement) but cannot increase the amount of work done! When using a lever, for example, it appears that we can lift incredibly heavy objects, such as a car with a car jack. Notice though that even using a hydraulic lift, we move the lever arm a great deal of distance, say 1 m, while the car only moves in height 1 cm. No matter the case : The work in can only equal the work out ideally and never actually does due to friction!

$$W_{in} = W_{out} + W_{friction}$$

$$Efficiency = \frac{W_{out}}{W_{in}}$$

All of these simple machines transfer energy and allow for changes from potential to kinetic, but there are different scales for this as well. Such as for electricity. Electrical current travels along conducting lines due to electrical potential. The higher the electrical potential, the further it can travel. However, there is friction of the electrons colliding with the atoms in the line. Collisions cause heating. This is called Joule heating, which in turn effects the efficiency of electrical power transfer. One of the critical areas in materials science in the last century and into today is the search for what are called high-temperature superconductors which would allow for electrical power transmission with little to no resistance in the line, hence little power loss.

What of this idea of Power?

Power

Thus far, we have noted the amount of Work done, but not the Rate at which the Work is done – that is the definition of Power. **Power** is the rate at which work is done.

$$P = \frac{W}{t}$$

The units are joule/sec which is called a **Watt**. This is the same unit we note in electrical devices, such as light bulbs. Two people can lift the same amount of weight when walking a flight of stairs, so they have done the same amount of work. Yet if one of them does it in half the time than the other then the one who takes less time uses more Power than the other. We can also convert the energy we use into horsepower, a common unit of power still in use today for engines and other mechanical devices.

This brings up the question of the differences between energy and power. Energy is simply related to a value to illustrate how much there is or needed for a situation, but it does not tell you how long it took for that energy to allow the phenomena in question to take place. For example, does it take more energy to climb a flight of stairs slowly or quickly?

This is a trick question! The answer is that both take the same amount of energy! We will do calculations to illustrate this in the activity. You might say that in one case you took ½ the amount of time to climb the stairs as compared to the other effort at it. What then is different about them? Only the amount of time is different. Since Work is the Change of Energy, recall the formula for Work. Work is a Force acting through a Displacement. Note that there is no mention of time. You can exert a force on a box and push it a distance across the floor. Independent of the speed, the amount of work done is the same for all cases. Returning to the stair-climbing in half of the time : the factor that changed here is not the work, it is the amount of time it takes to do it. There is a relation used in physics to examine Work per unit Time; this is called Power. (Power = Work per unit Time, $P = \frac{W}{t}$).

So a better question would be does it take more power to climb a flight of stairs slowly or quickly ? Obviously though they take the same amount of energy, the faster it is done, the more power is needed to accomplish the task. The rate of energy consumption is what makes one tired and causes us to heat up and hence a need to sweat in order to cool down.

Interesting to note is that the conservation of energy applies to power as well. The amount of energy per unit time going into a system, like a transformer which can alter the voltage, the power is the voltage times the current in the system, and this product going into the transformer and coming out will be the same. In essence the power going into a system always equal the power coming out of a given system. This too cannot be altered. (Hence $Power_{in} = Power_{out}$)

The ideas of Energy, Work, the Work-Energy Theorem, and Power are central to All Sciences and how they are looked at. In fact, how much energy is available can be found from a given piece of matter. Einstein's famous equation : $E = m*c^2$. This is the equivalence of matter and energy itself.

Consider the Weather. What drives the winds, the waves, and causes the temperature of the environment around us? **The Sun is the primary source of all energy on this planet for the most part.** The Sun heats the matter on the planet, the solids, liquids, and gases, and since different materials heat at different rates, and because the sun strikes these materials for different amounts of time at different angles, it results in differentiated heating, hence regions of hither and lower temperatures on the land, in the air and in the water. Also there are areas of different pressure in the air, convection cells of higher and lower pressures. These pressure differences result in our winds, which drive the waves too. Evaporated water becomes clouds which again fall as precipitation.

Geologic processes are driven by the stored energy of the Earth internally from radioactively decaying materials which heat and melt the interior regions of the Earth. This results in convection currents which in turn drive and move the plates of the Earth.

Biologically, all life depends on the Sun as well. The Sunlight is absorbed by plants through photosynthesis, which uses the energy, water, and carbon dioxide in a chemical process to produce sugars. This acts as the basis of energy and food for all food chains. The food we eat powers our bodies, for example.

Energy as we have noted is the central building block of all sciences – physics, astronomy, chemistry, biology, and any other science for that matter. We are as much a part of science as any other item in any other science is. Every action we do, walk, lift, even think, requires the use of energy. This, of course, is why we eat. The chemical bonds of the foods we consume contain energy in that their interactions in our bodies as we break them down allow the energy to transfer from those molecules to the molecules that make us up.

Look at how energy is oftentimes defined : "the capacity or capability to do work". The very equation for work is work is the change of energy of a system! $W = \Delta E$. What this means is that any and all actions and phenomena in nature require energy to happen. Energy cannot be created nor destroyed, it can only be transferred from one item to another and typically it changes from one form into another. We convert the chemical energy of our food into the kinetic energy of motion of our bodies.

Not only does the Sun power today through the foods taken in, but it was obviously shining long ago on plants then. Some died in shallow swamps and were buried over. Years of sediment piled atop them and chemical processing in the Earth has led to oil, natural gas, and coal deposits. Hence their name or term for them, the fossil fuels. These, in turn, have been excavated, mined, and pumped to be are used as power sources today. We burn coal in an electrical plant to heat water which in turn turns turbine blades that spin a generator (a system of magnets and wires) to generate electricity. Transformers change the voltage (increasing it, but therefore decreasing the current since the Power cannot be changed) to send it along power lines to our homes. Here transformers step down the voltage to usable levels in the home. In the house we plug in our appliances to use the electricity to create heat (for the toast in the toaster), light (from the light bulbs) , and sound (from the radio or TV) energy from burned coal that heated water to generate electricity. We can only use the energy available to do things, such as move electrons in wires.

As noted in the conservation of energy idea, energy cannot be created nor destroyed, only transformed and/or transferred.

This is true even for the seemingly inexhaustible energy source for many things, the Sun. First, a word on **Renewable** vs. **non-renewable Energy forms**. Actually given enough time and the right circumstances, energy forms like coal and oil can renew, but this takes millions of years to occur. What makes them non-renewable is the fact we are taking out far faster than their recovery rate. On the renewable energy side are things like solar, wind, and wave forms of energy. The wind and wave ones are driven by the sun and solar can be used as passive heating systems or active systems which tap the sun using solar cells to generate an electric current.

In any case, the Sun, as noted is not infinite either. But where does its energy stem from? Recall the aforementioned equation by Einstein equating mass and energy. In the core of the Sun some 5 billion tons of Hydrogen are turned into pure energy every second. The reaction actually has a byproduct too. The original amount in the reaction 500 million tons of Hydrogen is turned into 495 million tons of Helium. The difference, noted above, goes into energy and becomes the light and all other wavelengths given off by the Sun. Stars are fusion nuclear reactors with sufficient pressure (due to its mass and gravitational pull) and temperature (10s of millions of degrees) which turn low mass elements like Hydrogen into heavier ones like Helium and in the process give off energy. Humans are presently trying to tap into this and develop a fusion reactor which would solve most of the energy problems and needs of the world for many centuries, due to the fact that there would be no radioactive waste and the materials that can be used are sea water, which the Earth has a lot of.

Conclusion

To wrap up all of this talk on energy and bring it back to electricity – it is easy to see from the discussion that electricity does not simply appear from nowhere and requires work to enable it to function so that we can use it. This work means we must use energy resources that are available and not infinite to generate the needed electricity. The current majority of electricity we use comes from coal, oil, natural gas, nuclear materials, and dammed-up water that is allowed to pass by turbines to create electricity. In the aforementioned energy forms, clearly there is a lot of waste products that we are noticing an effect on the environment – i.e. global warming, plus they are not infinite. Much like the recycling philosophy of 'reduce, reuse, recycle' energy production and use needs to follow in these steps – to reduce our consumption, to increase efficiency, and in another direction to find other energy alternatives. There is a lot of talk and now efforts to move towards renewable forms of energy for electrical power generation including wind and solar. This also includes all the work in building a better battery for many reasons both large and small. These battery storage systems are going to be vital for alternative energy forms as well as massive changes in cars and mass transit systems. In the book there are not only Activities on electricity, electrical generation, but also wind and solar energy considerations.

Activity #1
Using Math (and the Slide Rule) in Everyday Life
Grade Level : Middle School
Math Level : Calculating

Everyday Life Calculations with the Slide Rule Activity

This Activity is the mathematical exploration of everyday life – such as determining miles per gallon, cost per unit ounce and the like, but all using a slide rule. Even if you don't, though I recommend it, these are basic things we should all know and engage in mentally to some degree. It is good to practice common math sense! :)

- o All of these calculations require either a question on the part of the person to consider the hypothetical or the actual items at hand to work on for calculations.
- o The materials needed depend on what is being done.
- o Know the basic rules for multiplication and division on a slide rule (they are summarized below and if needed in the first question – miles per gallon).
- o When scales other than C & D are used the rules are explained.
- o In most cases, only the C scale and D scales are needed.
- o Always estimate an answer and watch decimal placement.
- o Not always needed but is handy is a small pad of paper.

Slide Rule Basics

1) For almost all calculations in this Activity, the C & D scales are used.
2) If considering a Ratio or Fraction, It is best to see the C Scale as the Numerator since it sits atop the D Scale and the D Scale as the Denominator
3) To Multiply, Place the Right or Left Index as needed of a given scale, say the D Scale below the first number to be multiplied on the C Scale.
4) Next, read along the D scale to the other number to be multiplied by and read above it the answer on the C scale!
5) If the number is not available, use the other index (Right or Left) and start the process over as needed.
6) If Dividing simply place the Numerator value on the C Scale over the Denominator Value on the D scale.
7) Read the Answer opposite the Index of the D Scale on the C Scale.
8) What about the use of Scientific Notation?
9) It is best to consider all values in Scientific Notation in all calculations. This is done by setting the decimal point past the first non-zero number in the value under consideration and multiplying by 10 raised to the appropriate power.
10) The power is determined to be positive if the decimal needs to move to the right to obtain the original number (2000 is 2×10^3 for example), and negative if the decimal needs to move to the left to achieve the original number (0.002 is 2×10^{-3} for example).
11) When multiplying with values in Scientific Notation, simply add the exponents (Result = Operand 1 Exponent + Operand 2 Exponent).

12)

13) When dividing by Values in Scientific Notation, simply subtract the Denominator Exponent from the Numerator Exponent (Result = Numerator Exponent – Denominator Exponent).

14) Note the special case rules for linear slide rules :

15) If Dividing and the Slide Projects Right, Then Subtract One from Your Exponent Total to achieve the correct decimal placement.

16) If Multiplying and the Slide Projects Left, Then Add One to Your Exponent total to achieve the correct decimal placement.

17) Here is a full summary of decimal placement :

> **The Rules for Decimal Placement :**
> 1) Always first estimate the Answer.
> 2) Convert all values into Scientific Notation.
> 3) For Any Multiplied Values, Add up the Exponents and For Any Divided Values, Subtract the Exponents.
> 4) If for any single Multiplication operation the Slide moves to the Left, then add +1 to the exponent total
> 5) If for any single Division operation and the Slide moves to the Right, then -1 from the exponent total
> 6) Mentally treat any two values independently (just as whole numbers now) but keep track of what the exponent will be through the rules.
> 7) That is to say answer the question as to where the Slide has moved and factor in that addition or subtraction as needed for each operation independently.
> 8) Take this final answer and use the exponent figure arrived at to determine the decimal placement!

Miles per Gallon Measured

1) For this calculation it is best to begin with a full tank of gas. Once full write down the number of miles on the Odometer [called the starting miles] (or reset the trip odometer to zero).

2) Drive for some period of time (it can be a day, but if you have a regular work schedule it may be better to go for 2 or 3 days).

3) Needless to say you put more gas in some time later. At that time write down 2 numbers.

4) First the current reading of the Odometer now [called the ending miles] (or look at the trip odometer and write down the number of miles driven).

5) Second write down the exact amount of gas put in this time. This is the number of gallons used.

6) The 10" slide rule will give reasonable results despite the number of digits in your numbers, but it is best to round off as needed to facilitate the calculation.

7) If you did not use the trip odometer, determine the difference between your starting miles and your ending miles by subtracting them. This is your trip total.

8) Quick review for Division and Multiplication on a Slide Rule :

9) To Divide place the numerator read from the C scale over the denominator on the D scale and read the answer opposite the D scale index on the C scale.

10)
11) To Multiply place the left (right if needed) Index of the C scale over the first operand read from the D scale. Read along the C scale to the other operand and read opposite it on the D scale to find the Answer.
12) To determine the <u>average miles per gallon</u> :

$$\mathbf{Mpg} = \frac{\textbf{trip total}}{\textbf{number of gallons used}}$$

13) This value is important in many ways – On long trips and the car is gassed up, you can now estimate the range the car will go before filling up by *multiplying the Mpg by the number of gallons.*
14) On a map, knowing the distance to some place an estimate of the amount of gas needed can be determined by *dividing the Trip Miles by the Mpg.*
15) Also estimated costs can be found by taking *the number of gallons used and multiplying by the average cost of gas at that time.*
16) This calculation can go along with the Average Speed one as well.

Average Speed along with Distance and Time Determinations

1) To find the average speed what is needed is the total distance travelled and the total time taken for the trip.
2) This calculation can be done for cars, bikes, walking, boats, or any moving object under consideration.
3) If by car, either determine the total distance to be traveled by examining a map (paper for those who like to figure it out themselves or on the internet and a map system like Google Maps) OR
4) Actually travel the distance and either reset the trip odometer before the trip or write down the current odometer reading (starting miles) and then at the end of the trip write down the new odometer reading   (ending miles).
5) In either case of the trip, it must be traveled and times.
6) For the traveled trip be sure to not only keep track of miles but use a watch o⁻ a timer. On a watch mark down the start and end times.
7) Determine the miles traveled by subtracting the starting miles from the ending miles (Miles Traveled) and
8) Do the same for the amount of time to travel by subtracting the start time from the end time (Travel Time).

9) The Average Speed is found by :

$$\mathbf{v} = \frac{\textbf{Miles Traveled}}{\textbf{Travel Time}}$$

10) Note that one does not need the total miles and total times to determine an average speed for part of a trip.
11) If you are on a long trip and had the start time and start miles down, then at any point in the trip the average speed for that portion of the trip can be determined.

12)

13) This might be useful in the case of a very long drive where if the total miles to cover is known and the average speed is determined for some point in the trip, then the amount of time needed is found by :

14) First subtract the miles covered from the total miles. *Take the remainder and divide this value by the average speed. This will give the amount of time in hours left for the journey.*

15) Estimated times can be determined for an entire trip by first estimating the average speed and dividing it into the total miles of the trip. This gives the number of estimated hours for the trip.

16) If there is a situation where the average speed is known and the amount of time is also known then *the distance traveled is simply the average speed times the amount of time traveled.*

17) Always watch to see that units match up as needed! If not convert units to agree with the problem at hand.

18) A good exercise is to convert units for those who like math practice. For example, convert mph to miles per second or feet per second.

19)   Miles per Second = mph / 3600

20)   Feet per Second = Miles per Second * 5280

21) Also why not try the metric conversions as well, such as mph to km/hr and then convert these to m/s.

22)   Kilometers per Hour (km/hr)= 1.6(09) * Miles per Hour

23)   Meters per Second = (Km/hr) / 3.6

24) These ideas apply to taking a flight too. Know the distance flown and time the trip to determine the average speed of the plane.

25) Note that distances do not have to be miles, it can be any measurement unit. A bicycle can be looked at from the number of houses it passes in a given time, for example. A crawling insect might be done in centimeters/minute.

Cost per Unit Known (Mass, Volume, Weight, Item Number)

1. Most stores today will provide the cost of an item per unit ounce, per pound, per unit number of the item. The first question then to determine is this : Is it correct?
2. The next goal of this activity is to compare it to other brands or the same brand on the shelf only packaged in a different size.
3. Note that often the larger sizes are not labeled in the same manner of cost per unit as the smaller ones.
4. Be sure to read the labels carefully since the unit cost provided might be something like cents per ounce yet the material is labeled in dollars and pounds.
5. Therefore it is a good idea to keep in the back of one's mind basic conversions, such as :
- **1 pound = 16 ounces**
- **1 cup = 8 ounces**
- **1 gallon = 4 quarts**
- **1 quart = 2 pints**
- **1 pint = 16 ounces**
- **1 kilogram = 1000 grams**
- **1 kilogram = 2.2 lbs (pounds)**
- **1 ounce = 28.3 grams**

6. With these in mind it is best to choose base units for calculation and comparison, for example turn all units to ounces for mass or volume considerations.
7. Reading a label first determine the total number of ounces if not given.
8. To calculate the cost per unit ounce (CPO):

9. **Cost per Unit Ounce (CPO) $= \dfrac{\text{Cost (\$)}}{\text{No. of Ounces}}$**

10. For all other types of costs per unit whatever it is simply the cost divided by the number of ounces (number of pounds) (number of items) and the like.
11. In all cases these are rates and a rate is a ratio or fraction.
12. The best way to treat all ratios is to let the Numerator (here Cost) be the C scale and the Numerator be the D scale (here the number of Ounces). The answer is read opposite the D scale index on the C scale.
13. Once you know the cost per item it is easy to determine the total cost of a purchase by taking that ratio and multiplying by the number of items you intend to buy.

Computing Sales Tax or Determining the Tip

1) This activity is not a mistake though these two things are different circumstances. It turns out they are calculated in the same way on the slide rule.
2) For both you are mathematically taking a cost and adding to it a percentage of the cost.
3) Read the Cost on the C scale and place the left index of the D scale beneath it.
4) Now read along the D scale to the known sales tax rate or the desired amount of tip.

5)
6) Here each mark on the slide rule represents 1% or 1/100.
7) For example, The second mark is 2%, the fifth mark is 5%.
8) The easy mistake is misreading the D scale. Be sure to keep track of the decimal point here.
9) For example if the cost were $20 the first mark past the left index on the D scale has a reading of $20.20, the fifth mark is $21.00
10) Yet if the cost had been $200, then the first mark has a reading of $202 and the fifth mark has a reading of $210.
11) This illustrates the power of the slide rule since all numbers can be represented on this logarithmically-spaced number line and it is merely the decimal point that needs to be tracked.
12) When you read to 1.1 this is 10%. The value above it on the C scale is 10% greater than the original value.
13) If you want the amount of tax or tip you have to subtract the original cost from this new value to see how much the tip is.

Calculating Cost of a Sale Item (given percent off)

1) Many of us have been to stores proclaiming some given percentage off and large tables on cards describing what one pays.
2) With a slide rule this is a very easy task.
3) First, we need to acknowledge the trick in advertising. If it is 10% off, we are still paying 90% (100%-10%), and we pay 75% if it is 25% off (100%-25%) and so on.
4) Knowing this makes the task of finding our actual cost very easy and we can even find our savings too.
5) Read the Original Price on the C scale and place it over the Right Index of the D scale.
6) Now read from right to left along the C scale and find the percentage you are paying.
7) For example, if you have 10% off, read to 9 which is now considered 90% and read above it on the C scale your price.
8) If you have 25% off read to 7.5 since this is 75% and read the price above on the C scale.
9) What if you want the amount of savings? Instead of the right index, place the Price over the Left Index of the C scale.
10) Reading from left to right find the percentage off and find its value on the D scale above.
11) Note that you may have to use the Right or Left Index differently if the values go off the scale!
12) Notice the speed of this reading. In a single moment one can determine what any percentage is for any given value. Try that one with a calculator for those who think that faster!
13) By the way go back to the sales tax activity above if there is tax on it to determine the total cost if needed.

Computing Pay Amount from Rates of Pay

1) For estimated gross pay, take the rate of pay (R) on the C scale
2) Place this value over the index on the D scale
3) Read along the D to find the Number of Hours worked (H)
4) Find the answer on the C scale above (G)
5) Notes : First note that the pay rate has to be in the same time unit as is multiplied. Second, this is the gross without any withholding. To factor that in take an estimated percentage of pay withheld for taxes and use the percent off calculation to find an estimated Net Pay. Third, if there is overtime pay, be sure to calculate those hours at that pay rate and add that to the base gross pay.

G = R*H

Estimating the Electric (Water) Bill from Meter Readings

1) This activity here examines directly the process of reading the meter, taking the values there and finding an estimated bill (excluding taxes and other fees on the bill).
2) Depending on the meter you have for the Electric and Water Meters, this will affect how it is read.
3) There are some that give a direct reading of the current level of consumption, and if present then you can proceed with the calculation.
4) If the number is not directly on it, then you need to start on the first day of the month if you want to monitor it for the month. (If for a week, pick any day).
5) Read the dial carefully. Most have dials and these go in opposite directions with each dial, first clockwise, then counterclockwise, etc.
6) When the arrow is between numbers be clear on how to determine what value it is. The arrow is always the value it is coming from and until it reaches the next value on the dial.
7) If the meter is in a dial pattern, be sure to start with the largest place value and work your way around the dial for the other digits.
8) In both the Electric and Water Meter cases, you need the starting value and then an ending value some time later (keep track of the time between your readings – a month is a good time frame).
9) Simply take the Final Number you record and subtract the First Number you record. This Net Amount Used is what you consumed.
10) In the case of Electricity it is normally in kilowatt*hours (kWh) while in the case of Water it is 100s of cubic feet of water (Ccf).
11) Strangely, the Water Company typically charges for Ccf and not gallons. For those who want to know the gallons, take the Ccf number and multiply by 748 as there are 7.48 gallons in a cubic foot)
12) In both cases, one can call or look up on line the average cost per kWh or Ccf and then simply multiply your use by these values respectively to arrive at an estimated cost.

Cost = Net Amount Used x Rate

13) Note that this cost is estimated, since it does not factor in any type of tier-pay system for cost changes for different amounts, any taxes, fees, and other costs that may appear on the bill.

Home Project Needs for Painting and its Cost

1) The most basic of calculation is simple enough, first measure the Length (L) and Height (H) of a wall in a room in the chosen units (ft or m) and calculate the Area (A). (**A = L * H**)
2) To make the process simpler and avoid mistakes round all values up to the next whole number value. Most walls are 8 ft tall for example.
3) Be sure to sum up all of the walls. TA = total area (**TA = Sum of all Areas**)
4) Not Recommended , But For those who like some level of precision, in the case of inches divide the inches past the feet measure by 12 and tack this decimal value onto your feet measure, instead of the rounded up values used.
5) For still greater precision, you can subtract out non-painted areas, such as windows and doors (A = L*W) if you like.
6) Whatever the case, take the final number TA and divide by 300. This gives the number of gallons needed for rough, textured, or unpainted wallboard.
7) If it is smooth walls instead divide TA by 350.
8) Note that these are estimates. Always overestimate and round up. For example you are not going to buy 3.3 gallons, buy 4.
9) The final calculation is cost, simply take the Cost per gallon on C scale place over 1 on the D scale and read along the D scale to the needed number of gallons. The Cost is found on the C scale above.

Home Project Needs for Wallpapering and its Cost

1) The most basic of calculation is simple enough, first measure the Length (L) and Height (H) of a wall in a room in the chosen units (ft or m) and calculate the Area (A). (A = L * H)
2) To make the process simpler and avoid mistakes round all values up to the next whole number value. Most walls are 8 ft tall for example.
3) Be sure to sum up all of the walls. TA = total area (TA = Sum of all Areas) This is the Wallpapering Area.
4) For still greater precision, you can subtract out non-papered areas, such as windows and doors (A = L*W) if you like, instead of rounding up. Subtract the total of the non-papered areas from the papered ones
5) Use the chart below to find the Usable Yield Value. This is divided into the Wallpapering Area to determine the Number of Single Rolls Needed.

$$\textbf{No. of Single Rolls} = \frac{\textbf{Wallpapering Area}}{\textbf{Usable Yield}}$$

6) The final calculation is cost, simply take the Cost per roll on C scale place over 1 on the D scale and read along the D scale to the needed number of rolls. The Cost is found on the C scale above.

Pattern Repeat (Drop)	Usable Yield (American rolls)	Usable Yield (European rolls)
0 to 6 in.	32 sq ft.	25 sq ft.
7 to 12 in.	30 sq ft.	22 sq ft.
13 to 18 in.	27 sq ft.	20 sq ft.
19 to 23 in.	25 sq ft.	18 sq ft.

Home Project Needs for Carpeting and its Cost

1) The most basic of calculation is simple enough, first measure the Length (L) and Width (W) of the room in the chosen units (ft or m) and calculate the Area (A). (A = L * W)
2) To make the process simpler and avoid mistakes round all values up to the next whole number value.
3) *Not Recommended But For those who like some level of precision, in the case of inches divide the inches past the feet measure by 12 and tack this decimal value onto your feet measure, instead of rounding up.*
4) Determine the Number of Square Yards Needed by taking the Total Area (in square feet so far) and divide by 9.

Number of Yards Needed $= \dfrac{\textbf{Total Area in sq ft}}{\textbf{9}}$

5) Note that this Calculation is for Vinyl Flooring too! Follow the same procedure noted above down to the cost here below.
6) The final calculation is cost, simply take the Cost per yard on C scale place over 1 on the D scale and read along the D scale to the needed number of yards needed. The Cost is found on the C scale above.

Determining Recipe Needs through Proportions

1) These directions addresses the question of what to do when a recipe does not match your materials on hand and/or the question of how much is needed when you are changing the scale of the recipe to a larger or smaller yield.
2) This is the type of activity that the slide rule does very well since it is a proportion, which slide rules excel at.
3) The idea to solve this comes from noting that the ratio of the needed amount for a given ingredient in the recipe of a given number of servings will equal
4) The amount you need to have for your batch divided by the number of servings you wish to make!

$$\frac{\textbf{Recipe Requirement for Material}}{\textbf{Recipe Number of Servings}} = \frac{\textbf{The amount of Material Needed}}{\textbf{Number of Servings}}$$

5) This type of logic can be used for any and all variations of the same theme here. Remember to let the Numerator be the C Scale while the Denominator is the D Scale.
6) First place the known values of the Recipe over each other and look along the D Scale to the Number of Servings you wish to make and find the amount of the unknown needed material above. Be sure to employ the proper use of scientific notation when needed.
7) Keep in mind, you can determine how many servings you are making by looking in the opposite direction and going from what you have and looking back on the D Scale at how many servings it will make!

Basic Conversions :

1) The Slide Rule is the best conversion system around. This is because this operation is similar to the percentage calculation and the recipe calculation above.
2) Here, in this activity we can convert fractions to decimals or vice versa. We can also find equal fractions as well.
3) Also, if we need to convert one unit into another (inches into feet, feet into yards, ounces into pounds, inches into centimeters, and even complex ones like mph into kph) this is considered here.
4) In the case of fractions simply set the C Scale as the Numerator and the D Scale as the Denominator. Reading along the C Scale one finds all of the similar ratios.
5) (For example, if 2 on C is over 3 on D one finds 4 on C over 6 on D, and 6 on C is over 9 on D, et al. –
6) Not only that, but above the D Scale Index we find the decimal equivalent of the fraction, 0.66)
7) This means for any fraction or decimal we can find the other easily.
8) What is needed is first a ratio of what is known, say one unit over another. (For example 1 inch = 2.54 cm, 8 ounces = 1 cup, 1 foot = 12 inches)
9) Set up this ratio on the slide rule. It can be in either order where one is the C Scale value and the other is the D Scale value.
10) Now look for the other known value along the line that is known. On the opposite Scale will be the sought out answer.
11) (For example if Inches are on the C scale and Centimeters are on the D Scale and we wish to know how many centimeters are 4.5 inches, we read along the C Scale to 4.5 and find the answer below on the D Scale).
12) What this implies is that with a simple list of conversion factors, one can readily perform rapid calculations.
13) Conversions can include currency, one set of units into another, and the like.
14) For Currency conversions : 1 unit (say the dollar) equals N units of another currency. Set the Left Unit (1) over the equivalent on the D Scale. Reading along C is the number of dollars and below is the number of corresponding units in the other currency!
15) Another fact is this : Recall our calculations previously for miles per gallon, average speed, and cost per unit ounce.
16) Notice that each of these is a 3-Variable Formula where :

17)

18) What this means is that if the ratio is known (X & Y) then the answer is always found above the D Scale Index (N).

19) Also if we know the outcome value (N), we can estimate answers more readily.

20) (For example if our average speed is 45 mph, how long will it take to travel 280 miles. With 45 on C over 1 on D, read along C to 280 and below it on D is the answer : 6.2 hours)

21) This is how any and all 3-Variable equations can be treated!

22) Some of the basic set of 3-Variable functions needed are as follows :

$$\text{Rate} = \frac{\text{Cost}}{\text{\# Items, etc}} \qquad v = \frac{\Delta d}{\Delta t} \qquad a = \frac{\Delta v}{\Delta t} \qquad V = I*R \qquad P = V*I$$

23) We have used the first 2 here in this activity amongst others (area, etc) while the last 2 are Ohm's Law concerning voltage, current & resistance, and the final one is electrical power. The middle one is the formula definition of acceleration.

24) Note that there are many more and this is a small list. The Activities have these to learn from and use. Some involve the amount of food one eats and the graphical breakdown of these items. Another Activity calculates the amount of energy used and power exerted in exercising.

25) The key to this is your use of the Slide Rule as the means to connect math to the real world.

26) The Proportion Formula can be used in a number of cases (see other Activities involving its use) such as in the height of an object or distance to it.

27) For example if you know your height and measure the height of your shadow on a given sunny day, and measure the height of a shadow of another thing (tree, light pole, house) you can determine its height.

$$\frac{\text{Your Height}}{\text{Size of Your Shadow}} = \frac{\text{Height of Object}}{\text{Size of Shadow}}$$

Home Economics - Simple Interest on a Loan or Savings Account

1) This calculation is a quick look at the Basic Interest Equation and not the more complex one of being compounded daily or with some other time value.

2) Here, the Principal (P) is known, and perhaps the Interest Rate (R), along with the time period (T).

3) From these variables we can find the amount of accrued interest for a total time period (I).

4) This is a multiple step problem, so follow the directions.

5) First place P on the C Scale over the Index of the D Scale.

6) Read along the D Scale to R and find the answer to this on the C Scale, which we will call 'V'.

7) Now move 'V' to the D Scale Index and again read along the D Scale to the variable T.

8) Above T find the Answer I now on the C Scale.

9) Be sure to rearrange the formula as needed if solving for some other variable.

$$I = P*R*T$$

More Complex Loan Equation Calculations

1) This next excursion is the maximum use of math for this activity and requires concentration. Here we are going to examine the compounded interest rate on a loan or savings account.
2) First look at the equation :

$$A = P*(1+\frac{r}{n})^{n*t}$$

3) Here : P = Principal value, r = Annual nominal interest rate expressed as a decimal value, t = the number of years, n = number of times the interest is compounded per year, and A = Total Amount
4) All of the variables are known here except A. We want to determine the total amount owed or earned from the Principal at these given rates and conditions.
5) First compute with your slide rule r / n by using 'r' on the C Scale over 'n' on the D Scale. Call this X. Jot it down for reference.
6) Add 1 to X and multiply this by P on the slide rule.
7) For example place P on the C Scale over the D Scale Index and read along the D Scale to (1+X) and find the answer, Z, on the C Scale above.
8) Next multiply n*t on the slide rule. Call this M and jot it down.
9) Rewrite Z in Scientific Notation.
10) Reading Z (see step 7) on the D Scale, find the log (Z) on the L Scale. Recognize that this is the characteristic and the mantissa is the exponent from the Scientific Notation value. This new value is Q.
11) Now multiply this value, Q, by M (step 8) and find the value W.
12) Look up the characteristic (all of the numbers past the decimal) of W on the L Scale and find its corresponding value on the D Scale.
13) The decimal place is determined by the mantissa of W (the numeric value before the decimal).
14) With proper placement of the decimal, the value read on the D Scale is the Total Amount, A.
15) This takes time to do and to understand, but with patience, practice, and determination, you can succeed. As Slide Rulers say, Keep On Computing!

Activity #2
Calculating Cost of Electricity Usage at Home
Grade Level : Middle School
Math Level : Calculating

The Cost of Electricity Usage at Home with the Slide Rule

The first goal of the activity is to explore the units that are used to describe electrical power usage (the kilowatt*hour or kWh) and then to estimate the use of electrical power and calculate an estimate cost.

Let's start with Electricity. Many of us have heard terms like Volts, Amps, Watts, and many others, but how are these related to determining the electrical needs and use of a given appliance. The American home has 120 V AC source for all connections. In reality, most appliances are rated for 110 V to 120 V. The appliances we plug in are designed for this voltage, but through their electronics, they may use different amounts of current (I), or the Amps. It turns out that the product of Voltage and Current is Power. Power is measured in Watts. ($P = V*I$). So, if an appliance does not have its rating in Watts, but has the Amps listed, then one has to only multiply this by 120 V to find its power rating in Watts.

What is a Watt? Obviously from the formula it is a volt*amp, but what is that. Volt units are Joules per Coulomb while Amp units are Coulomb per Second. When canceled out correctly this leads to Joules per Second. Joules is an amount of energy and Seconds is the amount of time that this energy is transferred. So 1 Watt means that for every 1 second of operation there is a transfer of 1 Joule of energy. For example, a 75 W bulb is converting and using 75 Joules per second of operation! Recognize though that a Joule is a small amount of energy. Lift a cup of water weighing 1 N (about ¼ lb) 1m in 1s and this requires 1 J of energy, to illustrate the point.

The wattage is not the use of the item, however. We are not paying for electrical power, but for electrical energy usage. Energy is Power times Time. So in the case of Electrical Power Usage Cost comes from the Rate per kilowatt*hour times the number of kilowatts a device uses.

So the kilowatt *hour is a unit of energy, but how does it compare to other forms of energy? 1 kilowatt*hour of electricity is 3,412 BTUs or 860,369 calories. These latter units are NOT food calories, those are 1,000 of these calories listed. (How about other forms of energy? 1 cubic foot of natural gas has 1,000 BTUs or 252,164 calories of energy available).

Notice that it is kilowatts and not watts as well. This means for that 75 W bulb, we have to divide by 1000 first to make it 0.075 kW. If we run it for 10 hours then it would use 0.75 kWh. Typical costs range and depend on where one lives, the power source, cost for peak hour usage, and may also be in some sort of tier-cost system, but ranges from 12 cents to 30 cents per kilowatt*hour. If we take our bulb example and have a cost of 20 cents per kWh, then the 75 W bulb at 10 hours yields 15 cents of total cost.

Realize that that is one bulb and on for a short time, plus the fact that there are many other costs factored into bills such as taxes, fees and the like. Also for a given number used there may be one rate, but as usage increases the cost per unit time can be higher (a step up in price). The number of kilowatt*hours used has a wide range and depends on the usage of electrical items, their power and amount of time, plus the number of household members using them. It can go from several hundred kilowatt*hours in a month (typical home value used ranges 600 - 1100) to a couple thousand kilowatt*hours for large-scale use in the home and even thousands of kilowatt*hours use in large-scale business.

Of course, any operating electrical device contributes to the electrical costs, but we need to recognize that just because a label has its wattage on it, it does not mean you are using that amount when it is on. The label is rated at its maximum, so that a microwave on lower power settings will use less electricity for example. The opposite can be true as well. Speakers may say 40 W, but this is the rating of the sound energy output and not the power needed to generate this, which could be double this amount. Also there are small devices that run a lot even though not 'on', such as the electrical clocks on VCRs, DVD players, and others with displays.

The primary goal of this Activity is to monitor, measure, and calculate the usage and costs of electrical power consumption (which is similar to the water usage in that Activity). This activity extends from the everyday life calculations activity where the meters are read, and in that case, the total usage is known. Here it is the examination of the items that use the electrical power.

With the monitoring of life style, appliance usage, one becomes more conscious of usage and hence this can lead to considerations of reductions or using alternatives if applicable (solar panels, et al).

What is fun and fascinating about this activity is that it is not only personal, but can lead to surprises – maybe something uses more or less power, for example, than one previously thought. For example a 5 W night light on for 8 hours a day for 30 days uses 0.12 kWh. That amount of electrical power consumption is equal to a microwave using 1200 W for 6 minutes of time.

We compile lists of those items that use electricity and water and construct pie charts (and/or bar graphs) for comparison purposes.

What is important to note here is that you have parental permission and supervision. Do not move large electrical appliances, do not tamper with their plugs or wiring. In most cases you can find the information about a given appliance from the guides provided with it, on-line, or on a label on the device (only if it is accessible)

<u>Notes to Pre-Activity :</u>

1. The Electrical Activity has this idea in mind – monitor those things that use electricity for its usage. Keep in mind your safety and operate n a safe manner. Have parental permission and supervision.
2. For the Electricity Activity, the best device to use for common objects, like televisions, microwaves, computers, stereos, and refrigerators for determining the actual usage is a device called a Kill-A-Watt. It is plugged into the device and then it is plugged into the wall. Decide on the amount of time it will be plugged in for and monitor it regularly. Record the results in the table below.
3. If you do not have the Kill-A-Watt device, then monitor the amount of time a given device is running and use tables (such as on this Act vity or in the product's guide book or on-line) to determine the kilowatts that the device uses.
4. Obtain for Electrical Power consumption the rate(s) for kilowatt-hcurs used. Notice here that there may be more than one rate, what could happen is a tier-payment system where for a given range of usage, there is one cost, and it changes at greater amounts of usage.
5. Note that this activity can be considered an extension of the Everyday Life Calculations Activity where the reading of the Electric and Water Meters is done and the cost is calculated.
6. The key to both activities (Electrical Power Usage and Water Usage) here is timing. You do not necessarily need a stopwatch, but this can be used.
7. Recognize this only gives an estimate value for both items since you are not measuring all of the sources of use. Also the bills for each are based not only on consumption, but also taxes, fees, and other items.

Activity : Computing the Electrical Usage & Cost

Purpose : To estimate the cost and amount of usage of electrical power in a home during a month through measurement and calculation of kilowatt*hours for given appliances for cost determination.

Materials :

- Power Consumption Device, like Kill A Watt monitor,
- If no power consumption device, use Tables or Guides,
- Timer,
- Everyday home electric appliances,
- Slide Rule

Procedure :

1) Realize there are some items that you merely time after recording their wattage rating. This includes lights for example. You can also include any of the items in the table below. Note that this table is generalized and may not be your actual devices. Further research may be needed.
2) The major thing to do is to plug the Kill A Watt device into the wall and attach a given appliance, such as the refrigerator or television. Decide how long of a time it will be monitored. If for a week, realize you need to change that into the month by assuming that the usage for that time period will be constant for the other days. Rotate the device to other appliances that are not easy to monitor. – If using the Kill-A-Watt device, be sure to read the directions and follow them. It works with some items and not others.
3) Important Note : that the Kill A Watt will not work with high power appliances, like electric stoves, washers, dryers, and anything else of more than 120V. There are some power meters in the works for higher power ones, but none currently. Use the Table below for these ratings for cost estimates. In these cases, measure the amount of time it runs and multiply by the listed power rating.
4) If not using the Kill-A-Watt, then use tables-guides-internet information along with measured amounts of time the appliance is running to determine its power usage. Take for example the table given below on this Activity. You may have to estimate the power rating of your given device, since this table provides a range for some items.
5) Make a list as long as you want, but try to monitor the major appliances so that it reflects the greater electrical power usage.
6) Take all of your time measurements and turn them into hours.
7) Multiply the hours by the power rating (in kilowatts) for the item. This determines the Electrical Energy Usage.
8) Sum up the Electrical Energy Usage.
9) Determine the percentage and pie chart piece size for each appliance in your list.
10) Create either or both a Bar Graph and/or Pie Chart of Electrical Power Usage for each category of appliance.

11) This Electrical Energy Usage is then multiplied by the Cost (X cents per kilowatt) to find the cost.

12) Though the Slide Rule is a recommended tool, all of these calculations can be done with a regular or scientific calculator. Some scientific ones even have built-in averaging formulae. For those who like spreadsheets, the data can be typed in and the formulae then also be typed in its own cell where the formula references each of the measured variables in their respective cells, for example B1..BN has the measurements and values used in the equation while BN+1 has the formula for all of these variables (why not the A cells? Simple – use them to label you variables)

Data :

Appliance	Power Rating (W)	Time Used (hr)	Kilowatt*hours Total (kWh)	Total Cost ($)

Table of Common Power Ratings for Appliances :
Note : these are estimated ranges
For those with a range, you may have to do further research in your guide books or on the internet.

Appliance	Power Rating (W)
Electric Furnace	8,000 – 25,000
Electric Space Heater	600 – 1,200
Gas Furnace blower	750
Central Air Conditioner	3,500
Window AC	500 – 1,300
Ceiling Fan (speed & size)	25 – 100
Electric Clothes Dryer	4,400
Washer (cold/cold)	300
Washer (hot/cold)	2,800
Dishwasher (heated water)	3,600
Electric Oven	2,000
Desktop Computer & Monitor	150 – 340
Laptop Computer	45
Plasma TV (42″ – 56″)	100 – 500
LCD TV (32″ – 42″)	90 – 250
CRT TV (19″)	55 – 90
Microwave or 4-slot Toaster	1400
Coffee Maker	900
Range burner	800

Calculations :

Be sure to use your Slide Rule!

Time Usage :

$$t \ (hrs \ ) = \frac{Total \ minutes}{60}$$

Electrical Power :

$$P = V*I$$

Electrical Energy usage :

$W = \Delta E = P*t$
     (P in Watts, t in hours)

Cost of Electrical Energy Usage :

$C = \Delta E*R$
     (C = cost, E = Energy, R = Rate of electrical cost)

Graphing Formulae :

Decimal Value of Usage :

$D = \dfrac{\text{Part Amount}}{\text{Total Sum of All Parts}}$

Pie Chart Piece :

$P = 360°*D$

Conclusion :

Examine the Total Costs of a given appliance, its amount of usage, and consider are there alternatives or ways to reduce use. Also examine the pie chart and recognize where most of the electrical bill comes from and consider ways to decrease the areas of largest use.

Activity # 3
Estimating the Quantity of Electrostatic Charge
Grade Level : High School
Math Level : Challenging

Determining Charge with a Simple Electroscope Activity

The idea of a 'charged' object has been around for many centuries. The term 'electron' comes to modern times from ancient Greece where the ancient Greeks found that amber when rubbed with fur had the ability to attract bits and pieces of other fur, et al. The very term electricity comes from the Greek term for amber.

The problem with further investigation on this are mostly due to the fact that the realm of the atom is very small and electrons are even smaller, since they are part of an atom. In fact, even today, we may know the mass of the electron (9.11×10^{-31} kg) and its charge (1.6×10^{-19} C) but there is no known dimensions of it, and some models even treat the electron as a wave instead of a particle. The properties of mass and charge took until the early 1900s with the Milliken Oil Drop Experiment to find these values.

By the 1700s, notable persons, like Benjamin Franklin, were investigating the phenomena of electrified objects and their properties. Items, like the Leyden Jar were crafted and used to store electricity. Though called 'batteries' by Ben, today they are seen more as a Capacitor rather than a battery. The battery, however, also arose at these times with work by Volta. (Capacitor Activity #5, Battery Activities #'s 6,7,8,9)

One of the chief realizations was that there are two types of charge, called positive and negative, by none other than Ben himself. The next main realization came from Charles Augustin de Coulomb (where we come by the term, Coulomb's Force) who came to quantitatively determine that the amount of force between any two charges varies directly with the amount of charge and the force varies as the inverse-square of the distance separating these forces. He used a torsion balance with charged spheres to measure this effect.

$$F = \frac{k*q_1*q_2}{d^2}$$

F is the force, q_1 and q_2 are the charges in question and d is the distance between them.

Look at the structure of this formula. It is interestingly similar to the force due to gravitational attraction between any two masses as constructed by Newton over 100 years earlier than Coulomb's work. Each has a direct relation on a given property (the electrostatic force has charge, while the gravitational force has mass) and both are related to the distance by an inverse-square of the distance between the objects. (see Inverse-Square Law Activities for light and sound : #20 & #21)

When considering objects, we can readily divide them into two very general categories of materials : insulators and conductors (Note : There are items called semi-conductors, such as photovoltaic cells — which has its own Activity #27). In the case of insulators, though they are resistant to electrical charge flow, does not mean that charges cannot be deposited and built up on them. We all know this phenomena, such as running a comb through our hair on a dry day and having that 'static cling' effect, which can also be a part of clothes – especially coming from a dryer. We have played with this idea at some point in our childhood by rubbing a balloon in our hair and having it stick to a wall (what happens here is that the electrons now on the balloon as rubbed off our hair have a negative charge and induce a positive charge on the wall, since the somewhat mobile electrons on the surface are repelled and are driven away from that region that the balloon approaches – which now creates an attractive force between these opposing charges and leads to a temporary electrostatic bond), or better still scuffing our feet on carpeting and then touching an unsuspecting friend to cause a small shock.

In all cases of charge, it is the electrons that are transferred from one object to the next. The one that receives them builds up a negative charge while the one that loses them develops a positive charge. When combing our hair, electrons are transferred from us to the comb, hence our hair develops a temporary positive charge (all the more reason it is attracted to the comb as we pull it away).

The sum of the charges gained and the charges lost in these exchanges is zero. That is to say, **charge is conserved**. *This is one of the fundamental laws of nature of conservation.*

Some questions might arise in one's mind over these things, though. For a given insulator used as a charge receiver, how many electrons are on it? How large of a charge can be built up on it? (All objects have a maximum capacity – this is not explored here and when dealing with electricity, even in this static electricity lab, always exercise caution) How do the charges distribute over the surface? Does the shape or type of insulator affect the amount of charge that is on it? (Charges tend to have a certain amount of separation since they are the same charge, and the same charges repel each other – much like the idea that different charges attract each other) - Amongst many others.

The first question, though, that we are going to investigate is this : How much charge is on the object we have chosen to use in this case – the double pith ball electroscope? Despite its descriptive name, it essentially is two pieces of cork, Styrofoam, or some other insulating material suspended on two adjacent strings. On these we deposit a charge on from some other source and then take basic measurements of in order to find the quantity of charge on them. Knowing the amount of charge and the amount of charge on an electron, we can estimate the number of electrons on a given pith ball – something impossible to do for anyone before the early 1900s, despite the existence of simple desk-top objects like the electroscope since the time of Ben Franklin.

Our mathematics here is rather sophisticated and requires knowledge of vector analysis and some knowledge of trigonometry plus an understanding of Newton's Universal Law of Gravitation as well as Coulomb's Electrostatic Force Law. The formula is derived for you, but the preliminary steps are there so that one could do it for oneself as well. Also note, this Activity only uses charges built up on balloons, combs, and the like and has no connection to electrically conducting materials, electrical current, appliances and the like.

Purpose : To use a basic double pith-ball electroscope plus some measures of
Distance or angle and mass in order to determine the amount of
electrical charge and the number of electrons on each of the
pith balls.

Materials :

- Double Pith-Ball Electroscope (see Note & Photo),
- Ruler,
- Protractor,
- Balloon (or Comb or some other insulator that can hold a static charge),
- Mass Scale (that can read to 0.01 g – but see note in procedure if unavailable),
- Slide Rule

Note : One could buy one of these or they are easy to make by using items such
as Styrofoam peanuts as the pith balls. A better choice over regular string
is to use dental floss.

Photo :

Procedure :

1. First measure the length of the string attached to the pith balls. It is best to measure
   them to the same point on each one (the top, for example – though technically the
   middle is best since in Newtonian Mechanics, solid masses are treated as point masses
   acting through their center of mass) (L). Note, with two length values, take the
   average and use this as the length of the string.
2. The second measurement that needs to be taken is the total mass of the pith balls. The
   equation will use the average value for the mass and treat each as if they have the
   same mass.

3. Note : You will need a scale that can read to the nearest 0.01 g to have reasonably accurate results, but there is a way to determine the average mass, if you have your own pith balls.

4. If, for example, you are using Styrofoam peanuts, choose a number of them and cut them to the same shape with a pair of scissors. Take a large enough group of them to register a reading on the scale you are using. To determine the average value, take this total mass and divide by the number of peanuts on the scale.

5. With the average mass determined (m_{av}), now it is time to charge the electroscope.

6. Before doing so, you need to measure the equilibrium angle of the system. Use a ruler pointed away from the base of the system and choose a number (such as 4 cm or 2 inches) from which you will hold the protractor in an inverted fashion so that the vertex aligns with the strings of the pith balls and you can measure its equilibrium angle measure (Θe). Note the distance is done so that there is less chance of discharging or effecting the charged objects.

7. To charge the pith balls :

8. You can use a comb or a balloon – in either case pass it through your hair. Note : the drier the air and time of the year, the better the charge build up.

9. Deposit the static electric charge on both the pith balls. Try not to touch them as this will discharge them. You may have to do this a few times. You have succeeded when there is enough charge on them so that they repel each other and now have a new angle of separation to be measured.

10. Again hold the inverted protractor at the decided upon distance and measure the angle of apparent separation (Θs).

11. Determine the actual angle of separation of the pith balls (Θ) and record this result.

12. At this point you can discharge the system and begin again for other measurements. Note each set of measurements are a separate calculation.

13. Calculation steps :

14. You can choose to use the standard vector representation of the system and the basic formulae (and up to the intermediate steps) to derive your own equation for the charge on a pith ball – or simply use the one provided.

15. Note that this is a complex equation and will require several steps in using the slide rule, but can be done continuously. Like any algebraic equation, start from the inside and work it out.

16. Here it is best to start with the sine of the angle (first realize you have to determine this angle by dividing your angle of separation by 2. The sine of the angle is found on the C or D scale. The square of it is then found on the A or B scale.

17. Take this A/B scale value back to the D scale and multiply it by 4, then by L^2 (visually or with a cursor find what L^2 is if it cannot be done mentally) and finally by the tangent of half the angle of separation (this value, like sine is found on the C/D scale).

18. When done with all of that, divide by 9.

19. Through all of this be sure to keep track of the decimal values and the proper units used. L is measured in meters, for example. The constant 'k' has a power of 10^9. The sine and tangent values are decimal values.

20. Once a quotient is now determined, take the square root of this value by first finding it on the A/B scale and looking down onto the C/D scale. In this case, it is important to know where the decimal is since this will determine which side (left – odd (number of zeroes) or right – even (number of zeroes)) of the A/B scale is read.

21. Hint : Charges are very small things and will typically have a negative exponent.
22. You can now determine the number of electrons making up this charge by dividing your result by the charge on a single electron.
23. Note : Though the Slide Rule is a recommended tool, all of these calculations can be done with a regular or scientific calculator. Some scientific ones even have built-in averaging formulae. For those who like spreadsheets, the data can be typed in and the formulae then also be typed in its own cell where the formula references each of the measured variables in their respective cells, for example B1..BN has the measurements and values used in the equation while BN+1 has the formula for all of these variables (why not the A cells ? Simple – use it to label you variables)

Data :

Total Mass (m) of both pith balls : _____ (g)
Average Mass (m_{av}) of a pith ball : _____ (g)

Angle of Equilibrium (Θe) : _____ ($^\circ$)
Angle of Apparent Separation (Θs) : _____ ($^\circ$)

Angle (Θ) of charged & separated pith balls : _____ ($^\circ$)
Length of string from pith ball to vertex : _____ (m)

Depiction of Vectors involved :

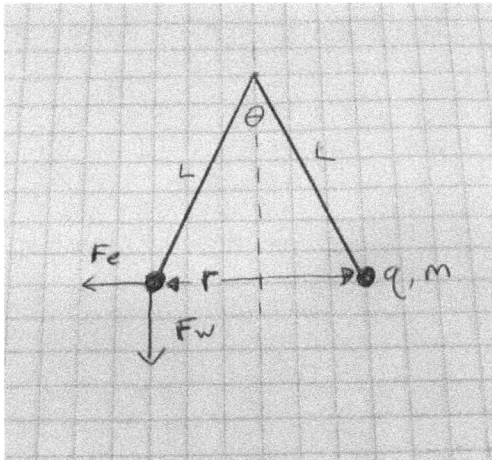

Calculations :

Be sure to use your Slide Rule!

Basic Formulae :

$\Sigma F_x = 0$
$\Sigma F_y = 0$

$F_w = m*g$

$F_E = \dfrac{k*Q_1*Q_2}{d^2}$

Intermediate Step :

$T*\sin(\frac{\theta}{2}) = F_E$

$T*\cos(\frac{\theta}{2}) = F_w$

$r = 2*L*\sin(\frac{\theta}{2})$

Angle of Separation :

$\Theta = \Theta s - \Theta e$

Derived Formulae :

$$Q = \left(\dfrac{4*L^2*m*g*\tan\frac{\theta}{2}*\sin^2(\frac{\theta}{2})}{k} \right)^{1/2}$$

Note : we are assuming that each of the pith balls has the same mass and the same charge

Average Mass :

$$m_{av} = \dfrac{\Sigma m}{n}$$

m : mass of a given piece, n : number of pieces

Constants & Conversion Factors :

$g = 9.8 \text{ m/s}^2$
$k = 9 \times 10^9 \text{ (N*m}^2)/(C^2)$
$1 \text{ kg} = 1000 \text{ g}$
charge on an electron : $1 \text{ e-} = 1.6 \times 10^{-19} \text{ C}$

Conclusion :

What do your results show? Does it seem like a large number of electrons? What should happen to the pith balls as more charge is added? Does the size of the pith balls affect the amount of charge that can be held on them?

Activity #4
Make and Measure a Leyden Jar Activity
Grade Level : High School
Math Level : Challenging

NOTE : Any and all Activities, such as this, involving the idea of electricity it is important to note the following : 1) You must have parental permission and supervision in conducting such activities, 2) Do not use electrical devices, appliances, outlets at all. This activity is about static electric charge build up and is not part of any electrical equipment 3) The following information is ONLY for historical reference to read and learn about ideas from and not an encouragement to do these historical experiments. 4) Always practice safety in your activities – do not do anything that poses a danger or health risk.

Historical Information about the Leyden Jar :

The Leyden Jar today would be best considered a Capacitor, despite thoughts that it seems like a battery. It essentially 'stores' static electrical charge between two electrodes on the inside and outside of a medium (the jar). The history, like many other objects in history, has two independent inventors : in 1744 a German cleric named Ewald Georg von Kleist (in fact some of these are called Kleistian Jars too) and sometime in 1744-46 by Dutch scientists Pieter van Musschenbroek of Leiden (Leyden, hence their name).

Leyden Jars were used in many early experiments in the new and arising field of electricity. The usefulness was primarily the compact size and it could store a large amount of static electrical charge.

Most jars were glass and had some sort of conducting metal foil surfaces on both the inside and outside of the jar. There is a gap distance between the exterior foil and the lid area so as to prevent discharging. Through a rubber, wooden, or glass lid projects an electrode that is rod-shaped that has often a chain attached to it and this chain is in contact with the inner foil surface. Some of the earliest ones used the medium of water in the jar.

Charging the Leyden jar is often done with an Electrostatic Generator in contact with the rod while the outer foil is grounded. With the development of charge, both the inner and outer surfaces will acquire equal, but opposite charges.

There is some famous history behind the Leyden Jar and experiments with them. Ben Franklin used many of them. He thought that the electric charge was stored not in the water, as many had assumed, but in the glass between the inner and outer plates. We know now this is not the case, and it is stored on the plates themselves.

Experiments showed that the thinner the glass medium (aka the dielectric), hence the closer the plates, along with other variables such as the larger the surface area of the jar, then there was a greater storage of static charge and hence a greater voltage. Later experiments showed that the dielectric material is not even essential, but increased storage capability (capacitance) and prevented any electrical discharge between the plates. Here is the connection to today's capacitor. By definition, any two plates separated by a small distance and having charge on their surface is a capacitor, even with no dielectric between them (or even a vacuum).

In fact, the amount of capacitance was actually measured in the number of 'jars' of a given size, or through the total coated area (assuming a standard thickness of the medium). A typical Leyden Jar that is one pint in size usually has a capacitance of about 1 nF. Capacitance is measured in Farads (named after an early experimenter in electricity) and is charge per unit volt ($C = \frac{Q}{V}$). The energy (measured in Joules) stored in a capacitor is equal to the work done to charge it. A Capacitor (formerly known as a Condenser) is used to store electric charge.

In time it was found that when connected in parallel this increased the voltage and total possible stored charge. They were put into use and appeared in many nations being investigated by scientists and engineers alike. Even as late as the early 1900s they were used in spark-gap radio transmitters and even medical electrotherapy equipment.

Today, capacitors are regularly used in electronic devices and are used to hold static charges. They can be used as timers or act in a sort of battery way. Capacitors are essentially plates separated by a gap, typically some sort of dielectric (i.e. insulator). With any potential difference across the plates, a static field develops across the dielectric and causes a positive charge to develop on one plate while the other develops an equal negative charge. Energy is then stored in the electrostatic field between the charges. They are often used to block direct current while allowing alternating current to pass. They can be used to smooth out the output of power supplies in filter networks. They are also used in resonant circuits that tune radios to particular frequencies (the tuner on our radios). (It is used in our Radio Activity # 24).

One of the other things that can be used with static discharge and developed by Franklin were the Franklin Bells. Franklin had a series of bells with a movable clapper between them. A wire running from a rod perched outside his home (the lightning rod, he developed) could have a charge induced on it by static in the air, which is common during a thunderstorm. This wire ran to the first bell while the second bell in line between the clapper for both of the bells was grounded by a wire. The induced charge would charge the first bell, which in turn induced a charge of opposite charge on the clapper and attract it, once in contact, some of that charge would transfer onto the clapper and hence repel the similarly charged bell and then strike the other bell transferring the residual charge to it, hence resetting then repeating the process. Franklin used these to act as indicators of impending thunderstorms.

Again Note that the history is merely for informational purposes and not for the Activity at hand in terms of what to do. Always practice safety, have parental permission and supervision in your work.

Our Activity will explore the Leyden Jar and find the voltage through knowing how sparks jump in standard air at regular temperatures and regular humidity.

Purpose : To construct an historic analog to the capacitor, a Leyden Jar, and determine how much charge is stored with variation in time of charging.

Purpose : To construct an historic analog to the capacitor, a Leyden Jar, and determine how much charge is stored with variation in the number or charging cycles.

Materials :
- Materials for the Electrophorus & Charging Materials :
- Disposable Aluminum Pie Tin Pan,
- Styrofoam Cup,
- Clear Tape (scotch tape – for all parts),
- Wool Cloth (sweater, socks, et al),
- Acrylic Sheet (size of piece of paper – large enough for pie pan),
- Wood Table,
- Materials for the Leyden Jar :
- Aluminum Foil,
- Small Plastic Jar (35 mm canister best),
- Thin Nut and Bolts (taller than jar),
- Nail (same diameter or less than Bolt),
- Thin insulated Wire (used here and later),
- OR thin Paper Clip,
- Other Materials & for Measuring the Charge :
- Wire Stripper,
- Small Piece Aluminum Sheet (cover to electrical box good or aluminum can lid),
- Styrofoam or poster board with foam center (sheet sized),
- Piece of Wire (18"),
- Ruler,
- Scissors or blade used in art,
- Plastic Soap Dish (rectangular),
- Graph Paper,
- Slide Rule

Set Up :

1) Assembling the Electrophorus :
2) Place an inverted Styrofoam cup in the middle of the aluminum pie tin pan and tape it down. – Set this aside –
3) Assembling the Leyden Jar :
4) Carefully push the nail through the lid of the 35 mm film canister (have parental help and exercise caution). Best to place it flat on a Styrofoam surface and push into it, instead of holding it.
5) Attach the bolt (or use the nail) to the lid – in the case of the bolt use a nut on either side of it.
6) Have the bolt project out a bit, but inside it does not touch the bottom but is mostly contained within an enclosed canister.
7) For the 35 mm canister carefully cut and line **both the inside and the outside of the canister** with aluminum foil. Make it as flat as possible. Adhere with minimal tape. Try not to have any gaps between the foil and the canister.
8) Attach a now stripped wire to the bolt inside the canister (i.e. under the lid) and place it so that when enclosed it touches the aluminum-foil lined interior.
9) Tools to measure the charge :
10) Stand upright in the horizontally-laying Styrofoam piece the piece of aluminum metal (best to use an electrical cover).
11) Mark with a marker a line parallel to the aluminum metal plate so as to have a place to measure from with a ruler (best to use metric).
12) Strip the ends of the wire and tape one end to the aluminum metal piece.
13) Place the completed leyden jar on its side in the soap dish bottom to see where the bolt (or nail) comes to. If needed (and with care) notch the soap dish with scissors or an exacto blade cutter so that the bolt juts out from the edge.

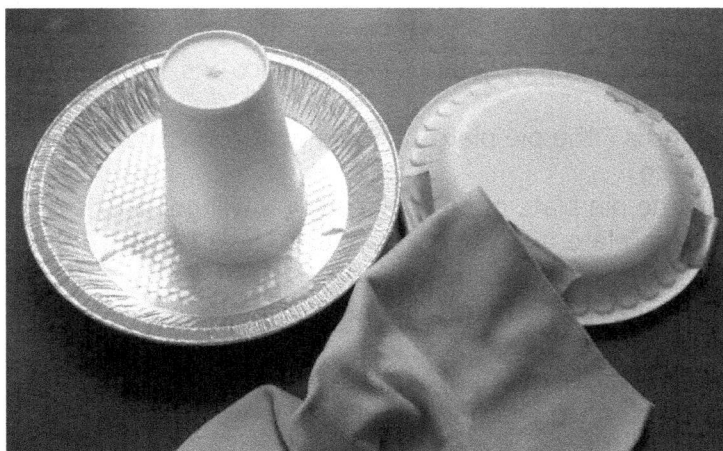

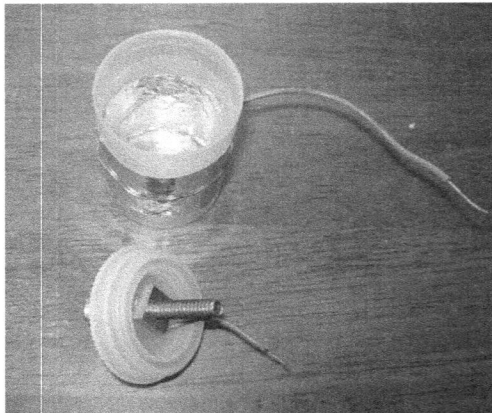

Procedure :

1) **Charging the Electrophorus :**

2) Be sure to decide ahead of time if you are testing the amount of time for a given set of cycles or a given number of cycles (hence all have the same amount of time) for the cycles of charging.

3) After clearing and cleaning the acrylic sheet piece (make sure there is no covering) place it on the wood table.

4) Rub the acrylic piece with your charging material (wool cloth). If the acrylic is new, do this for nearly 1 minute and discharge the static build up, then proceed with the process.

5) When rubbing the acrylic sheet with the wool material time yourself (see the data table), this is recorded in charging duration column. See the data table area for recommendations.

6) At this stage, the acrylic piece has a negative (excess electrons) charge built up on it.

7) In handling the electrophorus (the pie plate) handle by the Styrofoam cup. Place it squarely on the acrylic piece.

8) Briefly touch the edge of the pie plate once in place (you will get a little discharge). Repelled electrons in the pie plate now have a path and leave to the ground via you. Hence you leave the pie plate with a positive charge.

9) Between each of the experiments discharge the electrophorus.

10)   **Charging the Leyden Jar :**

11) Be sure to keep track of the order of Charge Cycles (have you done 1, 5, 10). For example, if 5 cycles, then you have charged the electrophorus 5 times and each time charged up the leyden jar each time, but now the leyden jar when measured for discharge gap distance will have been charged for 5 cycles.

12) Pick up the Electrophorus in one hand by the Styrofoam cup handle.

13) In the other hand hold the Leyden Jar with your hand on the outside aluminum foil.

14) Touch the Electrophorus pie plate edge to the bolt head. (Like in the case of touching the electrophorus you may hear a spark or zap).

15) Set aside the electrophorus.

16)

17) NOTE : When first doing the experiment, test the Leyden Jar by using a small bit of wire with ends stripped. Touch one end to the outside aluminum foil and bring it close to the bolt – you should hear and/or see a zap and a spark. – Once it is working skip this step.

18)   Measuring the Charged Leyden Jar :

19) Place the Leyden Jar on its side in the soap box and keep some distance between it and the aluminum metal plate (about 10 cm).

20) Attach the other end of the wire from the metal plate to the leyden jar and begin to slowly slide it close to the aluminum metal plate.

21) Now when the Leyden Jar discharges and you hear it you stop (note you have had to move it very carefully and slowly up to this point).

22)   Use a straight-edge to mark the place where the bolt-head is versus the aluminum plate so as to determine gap distance in the appropriate charge cycle column. (record this value). Measure this to the nearest $1/10^{th}$ of a mm.

23) Calculate Average Values for the Number of Charge Cycles in terms of charge gap distance. The more trials the better.

24) Determine the estimated Voltage of the Leyden Jar from the average values.

25) It is possible to see if there is a connection to the controlled variables here and have a rough estimate of voltage versus either number of discharge cycles or duration of charging by graphing this data and determining the slope.

Data :

Measuring the Voltage of the Charged Leyden Jar

Trial	Charging Duration (s)	Length of Spark from 1 charge cycle (d)	Length of Spark from 5 charge cycles (d)	Length of Spark from 10 charge cycles (d)
1				
2				
3				
Average				

Trial	Charging Duration (s)	Voltage from 1 charge cycle	Voltage from 5 charge cycles	Voltage from 10 charge cycles
Average				

NOTE : The Charging Duration is the amount of time that the wool is wiped across the acrylic sheet and should be the same for all 3 trials in a Session. The Duration periods may vary, but a suggestion is : 30s, 60s, 90s.

NOTE : One Charge Cycle is charging the Electrophorus and charging the Leyden Jar. This means that for 5 cycles, the Electrophorus and the Leyden Jar are charged 5 times for the duration period (30s, 60s, 90s). Pick a time when using cycles as your measure. If you are using a given number of cycles, then have a variation in time for the charging cycle.

NOTE : The Table has 1, 5, and 10 charging cycles. Each of these is its own experiment. First do 1, then 5, then 10 separately.

Calculations :
Be sure to use your Slide Rule!

Average Voltage :
R is the average voltage needed in air at 1 ATM for a spark length of 1cm

R = 30,000 V/cm

V = R*d

Note : 1 mm = 0.1 cm

Slope = $\frac{\Delta Y}{\Delta X}$

Conclusion :

What do your results show? If you were to try the experiment under different conditions (such as at different times of the year, with different levels of humidity) how do you think this may or may not affect charging the leyden jar? Do materials affect the charging of the leyden jar?

Activity #5
Variable Capacitor Creation and Measurement Activity
Grade Level : High School
Math Level : Challenging

Variable Capacitor Creation and Measurement Activity

Capacitors are mentioned in the Leyden Jar Activity (#4), but here is a summary of the key ideas about them. A Capacitor was once known as a Condenser and is a basically a 'passive' two-terminal electrical device which is used to store energy in an electric field.

The earliest capacitors were basically Leyden jars (see Leyden Jar Activity) developed as long ago as 1745 by Ewald von Kleist of Pomerania in Germany by storing charge in a glass jar with water connected by wire to an electrostatic generator. Though the earliest version, the Leyden Jars were named for Dutch physicist Pieter van Musschenbroeck while he worked at the University of Leiden. Even Benjamin Franklin experimented with them and helped coin the term 'battery' for them (though they are not). The term condenser was given by Volta who also worked with them in 1782, since he found they could store a larger amount or higher density of electric charge as compared to a normal and isolated conductor.

Essentially charge is developed (by some means) on one of the two plates in a given capacitor while the other plate has charge drawn away hence a potential difference (voltage) will develop across the plates. Between the plates is often a materials known as the dielectric. With the charges on the plates there is an electrostatic field, hence an electrical potential energy.

The capacitance of the capacitor is measured in the units of Farads which are coulomb/volt ($C = \frac{Q}{V}$). Capacitance is affected by plate size, the larger the capacitance. Also the gap between the plates affects this value. The smaller the gap, the greater the capacitance. Interestingly having a dielectric, though acting as a sort of insulation, actually increases capacitance at a given gap distance.

Capacitors come in a variety of sizes as well as materials that they are made of and hence have a wide range of capacitance (from a few picofarads (10^{-12}) to 5k Farads).

Capacitors are used for a variety of things. The first, and most obvious use, is since they act in a similar fashion to a battery, they are used to store energy. Here they help to maintain power supply in a given electronic device when main power batteries are off or out. They are used to block direct current while allowing alternating current to pass. They are used in what are called filter networks to smooth the output of power supplies. In the Radio Activity, we are using one where it operates in a resonant circuit which is used to tune radios to a particular frequency. Also in radio systems they are used to store energy for the amplifier for use when needed.

The types of materials, particularly in the dielectric, will affect not only the capacitance of the capacitor, but then also its use. Mylar dielectrics are often uses as timer circuits, such as in alarm clocks, regular clocks, and counters. A ceramic dielectric is uses for high frequency applications like those in antennas, as well as x-ray and MRI machines. Glass dielectrics are used in high voltage situations, while ones called super capacitors are used in powering electric and hybrid cars.

In our Activity, there are two explorations. In the first, we make a Flat Capacitor out of aluminum foil and cardboard and examine the property of surface area in relation to capacitance. We can also use this activity to explore the thickness of the dielectric as well, if we want. In the second activity, we make a Roll Capacitor and once again examine the property of capacitance with respect to the number of rolls (hence surface area) of the capacitor.

Purpose : To construct a capacitor and measure its capacitance by changing the surface area of the two capacitor plates (here aluminum foil) separated by a dielectric (here either paper or wax paper) to see the effect of surface area on capacitance.

Purpose : To construct a capacitor and measure its capacitance by changing the separation of the two capacitor plates (here aluminum foil) by increasing the thickness of the dielectric (here paper) to see the effect of plate separation on capacitance.

Materials :

- Aluminum Foil,
- Wax Paper,
- Thin Poster Board (comic backing boards work best),
- Scissors,
- Clear Tape,
- Multimeter that is capable of measuring capacitance,
- Pencil or Straw (used in capacitor construction),
- Graph Paper,
- Ruler,
- Caliper,
- Slide Rule

Note : Be sure to know how to set up and use your multimeter so that it measures capacitance. Also realize that if your capacitor is too large it may exceed the limits of the meter.

Note : In this Activity, there are 2 Activities – the first is for the Flat Capacitor and the second is for the Roll Capacitor.

Set Up Procedure :

1. **First : Flat Capacitor Construction**
2. Measure out two pieces of Aluminum Foil that are the same in size and will cover approximately 2/3 or more of the backing board. (see photo) Note that it is important to have the sizes known and recorded as these are part of the calculations.
3. If not using backing board, then measure the card board (if you want – or simply use the cardboard on the back of a pad of paper – note you need two of them the same size)
4. The average backing board for a comic is about 7" x 10".
5. With each of the pieces of Aluminum Foil it is best to have a 1 inch wide and about 2 inches long section that comes from the main rectangle as this will be the connection point for the multimeter. This part should hang over the edge of the cardboard.
6. In any case of a cardboard – aluminum foil system, tape the aluminum foil to the cardboard (see photos below).
7. It is best to have other pieces of cardboard IF you decide to test the effect of spacing between the plates (the aluminum foil covered pieces) with a dielectric (the blank piece in between the plates). This is a suggested Activity, the main activity explores the relationship of surface area to capacitance below.
8. **Second – Construction of the Roll Capacitor**
9. Here there are 4 sheets needed : 2 Aluminum Foil and 2 Wax Paper. They should be about the same size and can be cut into a rectangle where one edge is about 1-2 inches wider than a pencil or straw. (see photo below) Like the Flat Capacitor, it is best to have a measured size that is recorded and try to be consistent in the overlapping of the pieces as noted in the next directions where the displacement gap is the same so that calculation of covered area can be more readily done.
10. Begin with the wax paper on the bottom and then place alternating layers of aluminum foil and wax paper to the top layer of aluminum foil.
11. Notice that with each of the succeeding layers it is displaced about 1" to one side regularly (see photo).
12. This displacement makes it easy to determine which piece of aluminum foil is jutting out from the capacitor (one plate to the right and one to the left) for making connections to the multimeter for readings.
13. Note : Do not roll the system yet. This is actually done in the Activity itself. With each set of rolls (say 1 or 2 at a time, for example) you then would measure the capacitance of the capacitor and record your results all the way until the roll is complete.
14. Note : Other sizes and numbers of layers could be tested as well for comparison/contrast studies.

Pictures of Capacitor construction :

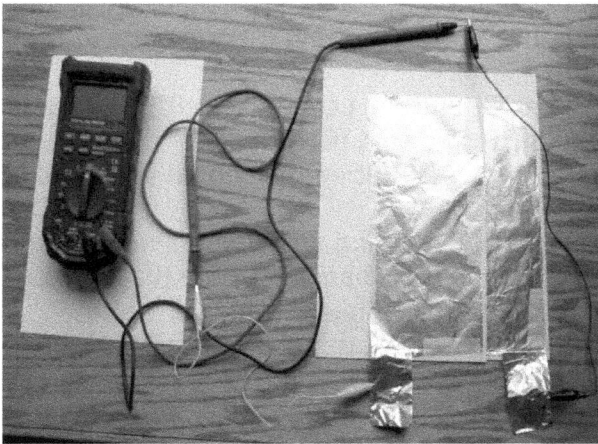

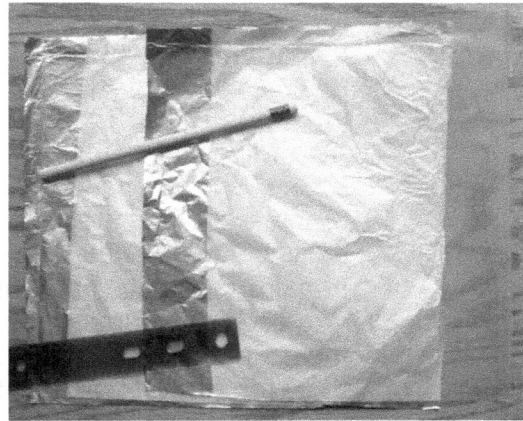

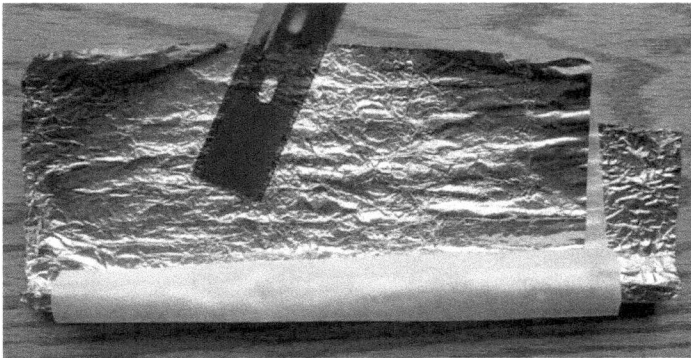

Procedure :

1. **Flat Capacitor Activity**
2. Once constructed, measure the Length (L) and Width (W) of the foil (they should be the same (excluding the tab piece)) and record these measurements in the table.
3. Also measure the thickness of a given plate (d) and record this – this id done with the caliper since it is so thin.
4. Decide on a strategy for how the two plates will cover each other.
5. For example, decide which direction will you move the top plate with respect to the bottom plate so that the areas covering each other varies with each move.
6. Note that it is best to not have the tab interact with the other plate, so slide the top plate so that the tabs never touch the other plate!
7. In terms of decision, you can determine the area of the plate and call this 100% and then decide that each trial represents another percentage, say 80%, then 60%, and so on to 0%.
8. For this to work, you have to recognize that one of the variables will remain constant, for example Width, hence you are moving the top piece in the dimension of Length.
9. As an example, if the Width were 10" and the Length were 8" the Area is 30 sq in. This is 100%. To move it to 80%, find that percentage of the total, which here will be 64 sq in. Since the Width is 10", then the Length value will have to be 6.4" and so on.
10. Another method is simple to choose decreasing areas and take down the measurements of the one dimension that is changing (note that the other remains constant) and from this the Area is determined.
11. For each given area, have the multimeter set up to read capacitance and measure the capacitance (C) of your capacitor and record this value on the table for the given area in the trial.
12. Graph each set of points where Area (can be Length or Percentage) on the x-axis and Capacitance (C) on the y-axis.
13. Draw the best fit line and determine slope for this situation.
14. If you want, use the basic equation for a Capacitor and determine how close your values approximate the constant in the formula ε, since you have measured C, A, and d.
15. In other experiments you can try different dielectrics. Use the same thickness cardboard and continue to place these between the plates and measure the capacitance. Use a caliper to measure thickness of this system and record this value and the independent variable (d).
16.
17. **Roll Capacitor Activity**
18. First construct the Roll Capacitor according to the instructions in the set up.
19. Each Trial is numbered and in the adjacent column in the Data Table is the number of rolls – it is best to choose a constant number for each measurement opportunity, such as 2 per each trial.
20. Measure the Roll Width (W) and the, the diameter of the pencil/straw used (d), and the thickness of the dielectric used (r_T).

21.
22. Use the formulae provided, create a table and calculate the diameter of the roll (r_o), length of winding of the capacitor (S), and determine the area of the capacitor (A)
23. In each trial measure the Capacitance and record this as well in the data table.
24. Graph the Capacitance (C) measured on the y-axis and the Number of Rolls (N) on the x-axis.
25. Draw a best fit line on the graph and determine slope to see if there is a relationship.

Data :

Activity 1 :

Trial	Thickness (cm)	Length (cm)	Width (cm)	Area (cm^2)	Percentage of Original Area (%)	Capacitance (C)
1					100	
2						
3						
4						

Activity 2 :

W : _____ (cm)
d : _____ (cm)
r_T : _____ (cm)

Trial	Roll Number (N)	Capacitance (C)
1		
2		
3		

Calculations :
Be sure to use your Slide Rule!

General Flat Capacitor Formula :

$$C = \frac{Q}{V} = \frac{\varepsilon * A}{d}$$

$$\varepsilon = \frac{d * C}{A}$$

Q : charge, V : voltage, A : surface area, d : separation, ε : constant 8.85×10^{-12} F/m

Flat Capacitor Formulae (to possibly use with General Flat formula):

Area of Plate formulae :

$A = L * W$

A : Area, L : Length, W : Width

Roll Capacitor Formulae :

Radius of Roll Capacitor :

$$r_o = \frac{d}{2}$$

r_o : radius of pencil or straw, d : pencil/straw diameter

Length of Winding in Roll Capacitor :

$$S = 2 * \pi * n * (r_o + r_T)$$

S : Length of Roll Capacitor, r_o : radius of pencil/straw,
r_T : thickness of wax paper and aluminum foil layer,
n : number of complete windings (1,2, etc)

Area of Roll Capacitor :

$$A = S * W$$

A : Area, S : Roll Length, W : Width of Roll

Graphing Formulae for all Capacitor forms :

Slope from lines graphed (Capacitance (nF) vs. Independent
Variable (either surface area (A)* or separation distance (d)) :
Note : * surface area depends on whether it is flat or rolled

$$m = \frac{\Delta y}{\Delta x} = \frac{\Delta \text{capacitance}}{\Delta \text{independent variable}}$$

Conclusion :

What sort of conclusions can you draw from your data and graphs
concerning the change in the amount of area and its effect on the
capacitance of the capacitor?

Activity #6
Factors Affecting a Simple Battery and the Slide Rule Activity
Grade Level : High School
Math Level : Calculating

History of the Battery :

The following is only about the history of the people and the events leading to the development of the batter. At no point are you to engage in these sort of activities or explorations. It is merely presented for historical reference and the operation of the basic battery. There are many ideas here involving chemistry and biology, some of which may seem unusual, even unethical today. This is only about information – and not to be done by anyone.

Following the history - Our Activity only explores a simplified battery composed of aluminum foil, pennies and salt water.

Today the world would find it very difficult, to say the least, and perhaps impossible to operate without batteries. We have them in our cell-phones, hand-held non-cord power tools, in our cars, used in numerous hand-held games and toys, in submarines, on the space shuttle, and in medical implants for our heart and ears. This is only a partial list as the number of devices can go on and on. Not only that, in the world today there is a greater need for them since there are efforts afoot to have vehicles use them as a power source itself.

Interestingly the term "battery" first comes to us from none other than Benjamin Franklin. He actually used the term to describe a series of interconnected Leyden Jars (which actually operate not as a battery, but as a charge storing device called today a Capacitor) since it resembled a battery of cannons. (In fact it was Franklin who gave us the terms negative and positive in referring to the charges of an atom). Today in the strictest of definitions the term, battery, refers to a collection of two or more cells, but through popular usage it also refers to a single electric cell.

There was a lot of electrical experiments in the 1700s, many of them done by Franklin amongst others. Here, the scientists of the day began to lay the foundations for later work in electricity of the mid to late 1800s which led to such innovations as the electric motor and electric dynamo, AC and DC systems, numerous electrical devices including the incandescent light bulb perfected by Thomas Edison and all of these ideas were fueled by Michael Faraday's Law of Induction (the connection of changing electric and magnetic fields) and the crafting of the Laws of Electromagnetism by James Clerk Maxwell.

Before all of these though there were names such as Luigi Galvani, an Italian anatomist and physiologist (where we today use his name as galvanic response) where in 1780 he notices that a dissected frog's leg would twitch in the presence of a spark from a Leyden Jar. Later in 1786 he noticed the same behavior during electrical or lightning storms. Further experimentation by him he tried two different metals with the leg to create an electrical circuit.

Alessandro Volta read of these accounts and decides that the frog's moist tissues can be replaced with cardboard soaked in salt water. Instead of a twitching leg to register response, he incorporates non-living things to respond to the electrical current by using a Galvanic Cell.

In the case of a given cell, the terminal voltage of a cell that is not discharging is called its electromotive force (emf) and has the same unit as electrical potential, named (voltage) and measured in volts, in honor of Volta.

In 1800, Volta makes the first battery by piling many voltaic cells in series. The larger the pile, the greater the emf. His pile would produce about 50 V for a 32-cell pile. There are still those that refer to the battery (in Europe) as **piles**.

Like many discoveries, the case and effect of the system was not fully understood for some time. Even Volta disagreed with many on the matter, for example, by wrongfully thinking that this was some sort of free inexhaustible energy. Today we know this is not the case and all things in the universe fall under the conservation of energy laws found to be the fabric of the cosmos.

All circuits require Voltage which is a measure of the Electrical potential energy per Coulomb (J/C) which is the definition of a Volt. This is to say that one volt is the electrical potential difference across which one coulomb of charge will gain or lose one joule of energy. This acts as the 'push' for the charges in a circuit. If the voltage is too low for a given circuit to operate, it will not. Common Voltage sources are DC (direct current) related such as dry cells and wet cells (the basis of this simplified activity presented here) and even extends to AC (alternating current) such as an electric generator (aka dynamo) and electric motors.

$$\Delta V = \frac{\Delta PE}{q}$$

It was **Michael Faraday** who found in 1834 that batteries were chemical reactions. According to Faraday, cations (positively charged ions) are attracted to the cathode and anions (negatively charged ions) are attracted to the anode.

These early batteries were great for experiments but not commercial development. Their voltages fluctuated and they could not produce a sustainable ample current for use. By 1836, what is called the Daniell cell was developed and even employed in industry for stationary items, such as telegraph networks. All of these initial batteries (often in glass) are what are called wet cells. In the late 1800s, a paste is made that operates in place of the wet environment and hence is born the dry cell battery. This makes portability much better and eliminates the fragility of the early wet cell system.

The electrical battery is a combination of materials which are electrochemical cells. These are used to convert stored chemical potential energy into electrical energy (which is often described as kinetic). Batteries consist of a number of **voltaic cells**, where each is composed of two half cells connected in series by a conductive electrolyte containing **anions** and **cations**. One half-cell includes electrolyte and the electrode to which anions (negatively charged ions) migrate (called the **anode** or negative electrode). The other half-cell includes electrolyte and the electrode to which cations (positively charged ions) migrate (called the **cathode** or positive electrode).

From chemistry, the battery is a **redox** (from oxidation-reduction) reaction. That is to say there is a reduction (addition of electrons, the reduction comes from a decrease in the oxidation number) which occurs to cations at the cathode, while oxidation (removal of electrons) occurs to anions at the anode. In the battery, the electrodes do not touch each other but are electrically connected by the electrolyte.

What this is saying and makes this occur is two metals that basically react with each other in the presence of the electrolyte. The two types of material, metals, in the voltaic pile are called electrodes. The reason for the potential difference is actually simple then, one of the metals reacts more vigorously than the other hence one metal becomes positively charged (the positive electrode) and the other becomes more negatively charged (the negative electrode). The potential difference is the voltage. This creates the pressure, force or push to allow current to flow when there is a complete circuit.

Each half cell has an **electromotive force** (emf) that drives electric current from the interior to the exterior of the cell. The net emf of the cell is the difference between the emfs of its half-cells (net emf = $E_2 - E_1$).

The electrical driving force or ΔV_{bat} across the terminals of a cell is known as the **Terminal Voltage (Difference)** and is measured in Volts. When it is not charging nor discharging, this value equals the emf of the cell and is called the open-circuit voltage. (The reason that the two values do not equal each other is that there is internal resistance in the battery, so that once connected the terminal voltage is a little less than the emf) [Note, for many labs in a regular physics class, internal resistance is neglected].

Despite differences in chemistry, there are categories of potential that have similarity for batteries. In one category, alkaline and carbon-zinc cells have approximately 1.5 V. In another set NiCd and NiMH (though different chemistries too) have 1.2 V. Finally, lithium cells can have emfs of 3 V or more.

There are two main classes of batteries : **Primary Batteries**, aka Disposable Batteries and **Secondary Batteries**, aka Rechargeable Batteries.

Disposable batteries are designed to be used once then discarded. Typically used in devices with low current needs with infrequent use. These cannot be recharged. Common forms of it are zinc-carbon and alkaline batteries.

Rechargeable batteries need to be charged before use and can be recharged by applying current to them to reverse the chemical reactions that occur from use. The oldest type of these is the lead-acid battery (used in cars often today). These types have large capacity (10Ah) and have peak currents of 450 A. Replacing many car ones in modern times is the sealed valve regulated lead acid (VRLA) battery. Some of these use a gel, or semi-solid electrolyte versus the older liquid acid ones or another type called the Absorbed Glass Mat form that has a fiberglass matting for absorbing the electrolyte.

There are a handful of Battery Cell Types which extends from the types of electrochemical cells that exist. The two most common will be discussed here : the **Wet Cell** and the **Dry Cell**.

The **Wet Cell Battery** (aka flooded cell, vented cell) has a liquid electrolyte. The other names comes from the fact that the electrolytic solution covers all the internal parts and these need to be vented since gases are produced during operation. These are the oldest major form of battery known. (Some of the names are : Daniell cell, Leclanche cell, Grove cell, Bunsen cell to name a few). Interestingly they may fall into either category of primary or secondary cells.

The **Dry Cell Battery** has a rather immobilized electrolyte that is in the form of a paste. There is some moisture but this is to allow current to flow. One of the distinct advantages of these is that they can operate in any position, unlike the wet cell (if tipped, it will spill the contents). Since the electrolyte is immobile this can lead to small batteries where there is little concern from leakage. The most common form of these type is the zinc-carbon battery (aka Leclanche cell) with a nominal voltage of 1.5 V which is the same as a alkaline battery. This is because each uses the same zinc-manganese dioxide combination. The standard makeup of the dry cell is a zinc anode (negative pole), usually in the form of a cylindrical pot, with a carbon cathode (positive pole) in the form of a central rod. The electrolyte is ammonium chloride in the form of a paste next to the zinc anode. The remainder of the space between the electrolyte and the carbon cathode is taken up by a second paste of the same material with manganese dioxide, where this latter material acts as a depolarizer. Modern 'high power' batteries replace the ammonium chloride with zinc chloride.

Description of the Activity :

This Activity is the construction of a simple battery and an examination of what will increase its voltage and overall power. We will use simple materials : pennies, nickels, aluminum foil, salt, water, paper towel, and the like for our construction. With a multimeter we will measure voltage and amperage of the battery we construct. Graph the results to see if there are changes with the changes to your experiment.

Activity : The Voltaic Penny-Foil-Pile Battery

Purpose : To construct a simple battery and measure its power with changes in the amounts of materials comprising the battery system.

Note for Safety : Always have adult supervision and permission for any activity. Do not involve outlets in any manner.

Materials :

- Multi-meter,
- 20 regular Pennies
- Sheet of Aluminum Foil (8" x 11"),
- 2 sheets Paper Towel,
- Dish Pan,
- Water,
- Salt,
- Scissors,
- Graph Paper,
- Slide Rule

Alternative or Extended Activity Materials :

- 20 Nickels,
- 20 Dimes,
- 20 older Pennies (pre 1982),
- ¼ c Vinegar,
- ¼ c Lemon Juice,
- ¼ c Soda Water,
- Thin Cardboard (from back of pad of paper),

Note : You need to have adult supervision and permission for this Activity. At no time are you to use any electrical device, regular batteries or any electrical outlet. The only tool that involves electricity is the multimeter, which is a galvanometer and responds to voltages and currents.

Procedure :

1) Rule No. 1 : See Note and be safe. This Activity is about the properties of materials, like aluminum foil, pennies, and salt water and does not involve electrical devices, regular batteries, and the like.
2) Prepare the electrolyte solution (water and salt) in a dishpan.
3) Side Note on Solution : Be willing to experiment with the concentration and perhaps instead try vinegar in a salt solution.
4) From the aluminum sheet, cut a 1" by 1" square and set aside.
5) Use the scissors to cut both the remaining aluminum foil and 1 sheet of paper towel into small squares the size of a penny. If possible cut as a circle instead to match the penny.
6) Side Note : You want as close a match as possible so that the layers are distirct and do not cross over as this may cause a 'short' in your system.
7) Soak, then lightly squeeze out the small 'squares' of paper towel.
8) Note : Take these pieces and place them on a easily tractable surface, like another sheet of aluminum foil for example for easy access.
9) Start the first pile with the aluminum foil on a flat table followed by the square (circle) of wet paper towel and then a penny.
10) This is your first pile.
11) Measure its voltage and current. With no readings, redo the experiment, or put on a second pile and try again. Measurement is done by using the multi-meter where one of the probes touches the aluminum foil and the other touches the penny on top of the stack of piles. Note that you have to rearrange the multi-meter to measure either in voltage or current.
12) Note : With multi-meters, the values will fluctuate a lot. Try to have one with a 'hold' button. Once the value seems to vary slightly around a given value, then hit the hold and accept that as the value.
13) Continue to create layers using the foil squares (circles) followed by the wet paper towel squares (circle) and the penny for each cell.
14) Measure after each cell is created both the voltage and the amperage and record the information in the table.
15) Once done, graph Voltage vs. Pile Number, draw a best fit line, determine slope.
16) Note : Determine the slope of the straightest part of the line, as it may vary to a curve possibly. For the more mathematically inclined : take the log of both the voltage and number of piles and graph these on a regular line graph. The slope of this line might reveal a power relation of voltage and number of piles.
17) Calculate the Power of each Pile. Note : Each of the reading is in milli- so this means 10^{-3}. When calculating Power in watts this needs to be considered.
18) Graph Power vs. Pile Number. Like the Voltage vs. Pile Number Graph, determine the slope of the straightest part of the line.
19) Try other variations listed below for comparison.

Data :

Note : A Voltaic Pile is : Penny – saturated towel – Foil

Number of Piles	Voltage (mV)	Current (mA)

Calculations :

Be sure to use your Slide Rule!

P = V*I

(P = Power, V = Voltage, I = Current)

Slope :

$$m = \frac{\Delta Y\text{-dependent variable}}{\Delta X\text{-independent variable}}$$

Conclusion :

Here the conclusion first comes from the calculation of Power for each of the Voltaic Piles created. How does this change? What did your graph illustrate to you? What of the change in Voltage with changes in the Pile size?

Side Note : One might notice that there seems to be no 'true' circuit when measuring current, which is often the process – that is to say, we should have some resistance or a load in place for our battery. We can insert a very low resistance (10 ohm) if we like between the multimeter and the battery, but from my observations of this test with and without one I find that it merely results in a smaller current value but does not affect its overall character. This is because there is resistance in the test wires already from the multimeter and they are quite low.

Extended Activities :

Each Activity is a Question to Explore :
1) From the Materials List, Try a different electrolyte concentration. We used 1 tbsp of salt to 6 cups of water. Now try 2 tbsp, then 3 tbsp to see how this affects the results.
2) From the Alternative Materials List, use a different electrolytic blotter. Instead of paper towel, use thin cardboard.
3) From the Alternative Materials List, use different types of piles : Penny & Nickel, Penny & Dime, Nickel & Dime, to test the outcome of the voltage and amperage of the battery produced.
4) From the Alternative Materials List, test new vs. old pennies (the older ones have a higher percentage of copper).
5) From the Alternative Materials List, use one or more of the other electrolytes, such as lemon juice, soda water, or vinegar and test how different electrolytes affect the outcome. Be sure to include distilled water as a control.

Activity #7
Factors Affecting a Coin Battery
Grade Level : High School
Math Level : Calculating

For History of the Battery see the Factors Affecting a Simple Battery Prelude.

This Activity is the construction of a simple battery and an examination of what will increase its voltage and overall power. We will use simple materials : pennies, nickels, salt, water, and the like for our construction. With a multimeter we will measure voltage and amperage of the battery we construct. Graph the results to see if there are changes with the changes to your experiment.

Activity : The Voltaic Coin Battery

Purpose : To construct a simple battery and measure its power with changes in the amounts of materials comprising the battery system.

Note for Safety : Always have adult supervision and permission for any activity. Do not involve outlets in any manner.

Materials :

- Multi-meter,
- 20 older Pennies (pre 1982),
- 20 Dimes,
- 20 Nickels,
- ¼ c Vinegar,
- ¼ c Soda Water,
- ¼ c Lemon Juice,
- Distilled Water,
- Medium Plastic Cups – This is the electrolyte solution that the cardboard (or paper towel) will be soaked in, - the number of cups is the number of tested solutions – be sure to include one for distilled water,
- Thin Cardboard (from the back of a pad of paper) – note : as in the simple battery Activity you can use paper towel instead,
- Scissors,
- Graph Paper,
- Slide Rule

Note : You need to have adult supervision and permission for this Activity. At no time are you to use any electrical device, regular batteries or any electrical outlet. The only tool that involves electricity is the multimeter, which is a galvanometer and responds to voltages and currents.

Note : Any one of the ¼ c liquids noted is useful — not all are needed for the Activity. Each is listed in order to test alternative ideas.

Procedure :

1. Rule No. 1 : See Note and be safe. This Activity is about the properties of materials, like pennies, nickels, vinegar (or other noted liquid in list) and does not involve electrical devices, regular batteries, and the like.
2. The first choice is to decide which of the electrolytes you will use (vinegar, lemon juice, et al). This may depend on what is available. If you have permission and materials availability you might be able to conduct this activity multiple times with each of the liquids for comparative purposes. In any case, this choice will be used for the entirety of the steps (#4-#21). Recognize that the goal is to test the coin combinations — this choice of solution is mostly for different options.
3. Even with the choice of the electrolyte, it is best to see how things operate without them, so first go through the steps using water (distilled water is best).
4. Next choice for each of the trials is the coin combinations. The first time around might be penny & dime, then the next trial might be penny & nickel, and then dime & nickel. You might even consider two of the same in a given trial just to see what happens.
5. The next step is the use of the material between the coins : Whether using cardboard or paper towel cut squared approximately 2 cm on a side (nearly and up to 1 inch if needed). Have as many as needed.
6. Have your electrolyte liquid (vinegar, lemon juice, et al) at hand. You will soak each of the pieces of cardboard (paper towel) as used. You might need to fish them out with a pair of tweezers. They do not need to soak long — you do not want them to degrade.
7. Have your coins sorted into stacks.
8. For the following directions I have chosen to illustrate it with penny & nickel — simply substitute whichever combinations you are using for this series of tests.
9. Start the first pile with the penny on a flat table followed by the soaked square of carboard or and then a nickel.
10. This is your first pile.
11. Measure its voltage and current. Be sure to know how to use the multimeter and its proper arrangement of testing probes and which setting (probably mV and mA). Run through its operation to be certain of how to do this.
12. If no readings, redo the experiment, or put on a second pile and try again. Note if using distilled water there may be cases of no readings.

13. Measurement is done by using the multi-meter where one of the probes touches the bottom of one coin (say the penny) and the other probe touches the other opposite side of the other coin (here the nickel) which is the top of the pile stack.

14. Note : With multi-meters, the values will fluctuate a lot. Try to have one with a 'hold' button. Once the value seems to vary slightly around a given value, then hit the hold and accept that as the value.

15. Now continue the test within a given electrolyte trial.

16. Continue to create layers using the 1st coin (here penny), square of soaked cardboard), then 2nd coin (here nickel).

17. With each pile, test the voltage and current readings and record these results in your table. (Note each pile can also be called a cell). Each set is a pile hence a number, the first is one, the next pile is two and so on.

18. Once done, graph Voltage vs. Pile Number, draw a best fit line, determine slope.

19. Note : Determine the slope of the straightest part of the line, as it may vary to a curve possibly. For the more mathematically inclined : take the log of both the voltage and number of piles and graph these on a regular line graph. The slope of this line might reveal a power relation of voltage and number of piles.

20. Calculate the Power of each Pile. Note : Each of the reading is in milli- so this means 10^{-3}. When calculating Power in watts this needs to be considered.

21. Graph Power vs. Pile Number. Like the Voltage vs. Pile Number Graph, determine the slope of the straightest part of the line.

22. Try other electrolytes and coin combinations for comparison.

Data :

Note (example): A Voltaic Pile is : Penny – saturated cardboard – Nickel

Number of Piles	Voltage (mV)	Current (mA)

Calculations :

Be sure to use your Slide Rule!

P = V*I

(P = Power, V = Voltage, I = Current)

Slope :

$$m = \frac{\Delta Y\text{-dependent variable}}{\Delta X\text{-independent variable}}$$

Conclusion :

Here the conclusion first comes from the calculation of Power for each of the Voltaic Piles created. How does this change? What did your graph illustrate to you? What of the change in Voltage with changes in the Pile size?

Side Note : One might notice that there seems to be no 'true' circuit when measuring current, which is often the process – that is to say, we should have some resistance or a load in place for our battery. We can insert a very low resistance (10 ohm) if we like between the multimeter and the battery, but from my observations of this test with and without one I find that it merely results in a smaller current value but does not affect its overall character. This is because there is resistance in the test wires already from the multimeter and they are quite low.

Activity #8
Factors Affecting a 'Lemon' Battery
Grade Level : High School
Math Level : Calculating

For History of the Battery see the Factors Affecting a Simple Battery Prelude.

This Activity is the construction of a simple battery and an examination of what will increase its voltage and overall power. We will use simple materials : copper wire (here it may be a good idea to rely on pennies, preferably older than 1982), steel wire (here really using a large paper clip), and the body of the battery will be one or more of the following (depending on your resources) : lemon, lime, banana, potato, or cucumber for our construction of a battery. With a multimeter we will measure voltage and amperage of the battery we construct. Graph the results to see if there are changes with the changes to your experiment.

Activity : The Voltaic Fruit/Vegetable Battery

Purpose : To construct a simple battery and measure its power with changes in the amounts of materials comprising the battery system.

Note for Safety : Always have adult supervision and permission for any activity. Do not involve outlets in any manner.

Materials :

- Multi-meter,
- Alligator clips on wires set,
- Pre-1982 Pennies or Copper Wire (18 gauge is good – needs to be stripped) – note : it is far easier and safer to use pennies,
- Large Metal Paper Clips,
- Wire Clippers (can have the wire stripper too) – Note : only needed if not using pennies,
- Fruit and/or Vegetables : Use older ones since these will not be consumed (never a good idea to consume materials used in a lab) : Choices can be items such as : Potato, Banana, Cucumber, Lime, Lemon, Orange (Note : Can have more than one of each of the one(s) chosen since we will test the idea of connecting batteries together in series too),
- Small Kitchen Knife,
- Graph Paper,
- Slide Rule

Note : You need to have adult supervision and permission for this Activity. At no time are you to use any electrical device, regular batteries or any electrical outlet. The only tool that involves electricity is the multimeter, which is a galvanometer and responds to voltages and currents.

Procedure :

1. Rule No. 1 : See Note and be safe. This Activity is about the properties of the fruits and/or vegetables we have chosen to make our battery and does not involve electrical devices, regular batteries, and the like.
2. Pre- Activity Preparation :
3. Note : The following Only needs to be done if using copper wire that needs to be stripped. It is suggested that pennies are the best alternative to this.
4. Note : It is best to have adults (parents) to do the preparation. The key is to be safe.
5. If possible in buying copper wire that is bare. If not possible, then you have to strip the pieces you are going to use. Each piece will be 2-3 inches (5-8 cm). It is probably easiest to strip a small stretch of wire from the main coil and then cut off a bare piece at the appropriate length.
6. Note : The number of pieces will depend on how many batteries you create. Each piece of fruit or vegetable used is a cell or a battery and each will need one piece of copper wire. A good Activity might be up to 4 at most – resources permitting (but can be fewer or more if interested).

7. The Set up of the Batter in the Activity :

8. As it is not known what item(s) you are using for your Activity (lemon, potato and the like) I will write this from the point of view of a lemon being the main item. You simply insert what you are using in place of the term Lemon.
9. Also I will assume that you are using pennies and large paper clips.
10. The Lemon, one penny, and one large paper clip are all that is needed.
11. Note that each piece of fruit solutions is a Trial.
12. If possible It is best to have at least 2 or 3 pieces of fruit (the Lemon) to make batteries with. There is another low cost solution, though. A given lemon can be cut in half, hence acting as two batteries!
13. The Battery :
14. Before cutting a Lemon so that it can be two batteries, roll it firmly on the table (not so hard as to rupture the skin of the Lemon).
15. Have your parents cut the Lemon so that it can sit flatly on the table (this can be done by cutting it in half and then cutting off each of the ends).
16. With the first piece insert into the actual exposed Lemon the penny and one of the paper clips. Have some distance between them (about 1-2 cm is sufficient).
17. In inserting the penny and paper clip leave enough out so that the multimeter can be used to touch them in order to take measurements.
18. Amazingly this is the first battery! :)
19. With sufficient materials set up two or three more if you are able to.

20. Measurements in the Activity

21. Be sure to know how to use the multimeter and its proper arrangement of testing probes and which setting (probably mV and mA). Run through its operation to be certain of how to do this.
22. We are going to measure each of the batteries separately (give them names or numbers to distinguish them in the table of data).
23. Measurement is done by using the multi-meter where one of the probes touches the penny and the other probe touches the paper clip in the same Lemon.
24. Record these values voltages and currents in the table for each of the batteries.
25. Now With the individual measurements done, we are now going to look at connecting our batteries in series and see how this affects the outcome.

26. Series connection :

27. Presently all the batteries are in a row where one of the poles is on the left (for example the penny) while the other is on the right (the paper clip following our example). – Be patient, the pennies are probably the hardest to connect.
28. Using the alligator clip wires connect the paper clip of the first battery to the penny of the next second battery.
29. The next alligator clip now goes on this second battery on the paper clip.
30. These two batteries are now in series and we can measure the voltage and current of two batteries and place these values in our data table.
31. To measure these batteries in series one probe of the multimeter touches the penny of the first battery and the other probe touches the paper clip of the second battery.
32. Now, leaving the first two batteries connected now put a third battery in series by attaching the paper clip of the second battery to the penny of the third battery.
33. Measure the data for this set of batteries. Continue this for all of your batteries so that all end up in series. Each new one results in two new measurements of the total voltage and total current of these batteries. Recognize that you always place the probes on the open battery posts (that is there is no alligator clips on them).
34. Note : Once done the first battery has a battery with no alligator clip and the last battery has paper clip with no alligator clip.

35. Calculations :

36. Once done with all of the measurements first Calculate Power for each of the batteries individually. Note : Each of the reading is probably in milli- so this means 10^{-3}. When calculating Power in watts this needs to be considered.
37. For the series connection : For each series (2 batteries, 3 batteries, et al) calculate the Power of this set of batteries.
38. In the series connection : Graph Voltage vs. Number of Batteries in series. Draw a best fit line and determine the slope of it.
39. In the series connection : Graph Current vs. Number of Batteries in series. Draw a best fit line and determine the slope of it.
40. Try other pieces of fruit and/or vegetables for comparison.

Data :

Battery Name or Number	Voltage (mV)	Current (mA)

Number of Batteries in Series	Voltage (mV)	Current (mA)

Calculations :

Be sure to use your Slide Rule!

P = V*I

(P = Power, V = Voltage, I = Current)

Slope :

$$m = \frac{\Delta Y\text{-dependent variable}}{\Delta X\text{-independent variable}}$$

Conclusion :

Here the conclusion first comes from the calculation of Power for each of the Batteries created. How does this change? What did your graph illustrate to you? Did you try other pieces of fruit and/or vegetables and how do each of these compare? If you were able to create a series of lemon batteries what does your graph show you in terms of more and more batteries in series with each other?

Side Note : One might notice that there seems to be no 'true' circuit when measuring current, which is often the process – that is to say, we should have some resistance or a load in place for our battery. We can insert a very low resistance (10 ohm) if we like between the multimeter and the battery, but from my observations of this test with and without one I find that it merely results in a smaller current value but does not affect its overall character. This is because there is resistance in the test wires already from the multimeter and they are quite low.

Activity #9
Factors Affecting a Solution Battery
Grade Level : High School
Math Level : Calculating

For History of the Battery see the Factors Affecting a Simple Battery Prelude.
This Activity is the construction of a simple battery and an examination of what will increase its voltage and overall power. We will use simple materials : copper wire, steel wire (here really using a large paper clip), salt, water, possibly other liquids such as lemon juice, vinegar for our construction of a battery.
With a multimeter we will measure voltage and amperage of the battery we construct. Graph the results to see if there are changes with the changes to your experiment.

Activity : The Voltaic Solution Battery

Purpose : To construct a simple battery and measure its power with changes in the amounts of materials comprising the battery system.

Note for Safety : Always have adult supervision and permission for any activity. Do not involve outlets in any manner.

Materials :

- Multi-meter,
- Alligator clips on wires set,
- Copper Wire (18 gauge is good – needs to be stripped),
- Large Metal Paper Clips,
- Wire Clippers (can have the wire stripper too),
- Small or Medium Plastic Cups,
- Clear Tape,
- ¼ cup,
- Tablespoon,
- Vinegar and/or Soda Water and/or Lemon Juice and/or cola pop,
- Distilled Water,
- Salt,
- Graph Paper,
- Goggles,
- Slide Rule

Note : You need to have adult supervision and permission for this Activity. At no time are you to use any electrical device, regular batteries or any electrical outlet. The only tool that involves electricity is the multimeter, which is a galvanometer and responds to voltages and currents.
Note : Any one of the liquids noted is useful – not all are needed for the Activity. Each is listed in order to test alternative ideas.

Procedure :

1. Rule No. 1 : See Note and be safe. This Activity is about the properties of the liquids we have chosen to make our battery and does not involve electrical devices, regular batteries, and the like.

2. Pre- Activity Preparation :
3. Note : It is best to have adults (parents) to do the preparation. The key is to be safe.
4. If possible in buying copper wire that is bare. If not possible, then you have to strip the pieces you are going to use. Each piece will be 2-3 inches (5-8 cm). It is probably easiest to strip a small stretch of wire from the main coil and then cut off a bare piece at the appropriate length.
5. Note : Length of wire is mostly connected to the size of the cups used. The smaller the better, hence shorter wire pieces.
6. Note : The number of pieces will depend on how many batteries you create. Each cup used is a cell or a battery and each will need one piece of copper wire. A good Activity might be 4 (can be fewer or more if interested).

7. The Set up of the Battery in the Activity :

8. The first choice is to decide the order of the electrolytes you will use (vinegar, lemon juice, et al). This order may depend on what is available. If you have permission and materials availability you might be able to conduct this activity multiple times with each of the liquids for comparative purposes. In any case, this choice will be used for the entirety of the steps (# 12 - # 44).
9. Even with the choice of the electrolyte, it is best to see how things operate without them, so first go through the steps using water (distilled water is best).
10. The next best set electrolyte is water with salt. It is best to use a pitcher to mix 1 tbsp to each quart of water for the initial salt water battery trial. Note you can continue the salt water trials and simply increase the salt concentration by adding more salt (after first refilling the water to the starting line) to each trial solution.
11. So a good selection might be : distilled water, salt water (with 1 tbsp), salt water (with 2 tbsp), then perhaps lemon juice or vinegar. Further studies might consider these latter ones with salt or instead still other trials with water with more salt.
12. Note that each of these solutions is a Trial. It is best to have at least 3 or 4 cups with the solution in them. All have the same amount of liquid (use a measuring cup to do this).
13. To the same side of each cup (now placed in a row) with the solutions one of the large paper clips is attached and pushed down to a point to be in the liquid (fill cup as needed).
14. On the opposite of each cup one of the bare copper wire segments is taped. So each cup is filled with a given solution and attached to the rim on each side is the wire piece and the paper clip. Each of these is a battery or cell.

15. Measurements in the Activity

16. Be sure to know how to use the multimeter and its proper arrangement of testing probes and which setting (probably mV and mA). Run through its operation to be certain of how to do this.

17. We are going to measure each of the batteries separately (give them names or numbers to distinguish them in the table of data).

18. Measurement is done by using the multi-meter where one of the probes touches the bare copper wire and the other probe touches the paper clip in the same cup. − Record all of these in the data table for each battery

19. With the individual measurements done, we are going to look at connecting our batteries in series and see how this affects the outcome.

20. Series connection :

21. Presently all the batteries are in a row where one of the poles is on the left (for example the copper wire) while the other is on the right (the paper clip following our example).

22. Using the alligator clip wires connect the paper clip of the first battery to the copper wire of the next second battery.

23. The next alligator clip now goes on this second battery on the paper clip.

24. These two batteries are now in series and we can measure the voltage and current of two batteries and place these values in our data table.

25. To measure these batteries in series one probe of the multimeter touches the bare copper wire of the first battery and the other probe touches the paper clip of the second battery.

26. Now, leaving the first two batteries connected now put a third battery in series by attaching the paper clip of the second battery to the copper wire of the third battery.

27. Measure the data for this set of batteries. Continue this for all of your batteries so that all end up in series. Each new one results in two new measurements of the total voltage and total current of these batteries. Recognize that you always place the probes on the open battery posts (that is there is no alligator clips on them).

28. Note : Once done the first battery has a copper wire with no alligator clip and the last battery has paper clip with no alligator clip.

29. Calculations :

30. Once done with all of the measurements first Calculate Power for each of the batteries individually. Note : Each of the reading is probably in milli- so this means 10^{-3}. When calculating Power in watts this needs to be considered.

31. If you have multiple sets of data for salt water solutions (that is − one tbsp, two tbsp, et al) then graph Voltage vs. Tbsp Number, draw a best fit line, determine slope.

32.
33. Note : Determine the slope of the straightest part of the line, as it may vary to a curve possibly. For the more mathematically inclined : take the log of both the voltage and tablespoon number and graph these on a regular line graph. The slope of this line might reveal a power relation of voltage and the tablespoon number (ie salt concentration).
34. For the series connection : For each series (2 batteries, 3 batteries, et al) calculate the Power of this set of batteries.
35. In the series connection : Graph Voltage vs. Number of Batteries in series. Draw a best fit line and determine the slope of it.
36. In the series connection : Graph Current vs. Number of Batteries in series. Draw a best fit line and determine the slope of it.
37. Try other electrolytes and/or concentrations for comparison.

Data :

Battery Name or Number	Voltage (mV)	Current (mA)

Number of Batteries in Series	Voltage (mV)	Current (mA)

Calculations :

Be sure to use your Slide Rule!

P = V*I

(P = Power, V = Voltage, I = Current)

Slope :

$$m = \frac{\Delta Y\text{-dependent variable}}{\Delta X\text{-independent variable}}$$

Conclusion :

Here the conclusion first comes from the calculation of Power for each of the Batteries created. How does this change? Which solution worked best? What did your graph illustrate to you? What of the change in Voltage with changes in the Number of Batteries connected in series?

Side Note : One might notice that there seems to be no 'true' circuit when measuring current, which is often the process – that is to say, we should have some resistance or a load in place for our battery. We can insert a very low resistance (10 ohm) if we like between the multimeter and the battery, but from my observations of this test with and without one I find that it merely results in a smaller current value but does not affect its overall character. This is because there is resistance in the test wires already from the multimeter and they are quite low.

Activity #10
Ohm's Law Activity
Grade Level : High School
Math Level : Challenging

The basic circuit is simply composed of a voltage source and a load. The load can be very complex with resistors, transistors, capacitors, inductors, and all sorts of other elements. These too can be connected in more than one way, the most common being either connected in series or in parallel to the voltage source as well as to each other.

The place to begin is the basic circuit. A **Circuit** is any complete path along which charge can flow. If there is a break in the circuit, the system stops.

The charges that flow are the electrons, but this is not defined as conventional current. **Current** itself is the flow of charge measured in Amperes (coulombs per second), and conventional current is described as the flow of positive charge. The symbol for current is I. The rate of current flow is directly related to the voltage of the circuit and inversely related to the resistance of the components of the circuit.

$$I = \frac{\Delta q}{t}$$

All circuits require **Voltage** which is a measure of the Electrical potential energy per Coulomb (J/C) which is the definition of a Volt. This is to say that one volt is the electrical potential difference across which one coulomb of charge will gain or lose one joule of energy. This acts as the 'push' for the charges in a circuit. If the voltage is too low for a given circuit to operate, it will not. Common Voltage sources are DC (direct current) related such as dry cells and wet cells or AC (alternating current) such as an electric generator (aka dynamo).

$$\Delta V = \frac{\Delta PE}{q}$$

At first it might seem that if there is any resistance in a circuit it would be considered an undesired thing, since by definition resistance is an impedance to the flow of electricity. The resistance is often put in to control the flow of the electricity, hence the use of items called resistors which offer a measured amount of resistance to the amount of voltage being used hence allowing only a certain amount of current in that situation. If the relation for the resistor varies directly with the voltage and inversely with the current, then the resistor is said to perform in an Ohmic fashion, that is obeying Ohm's Law. These components are represented by a zig-zag line symbol and have a given value, measured in Ohms of resistance. An Ohm is really a volt per amp.

Ohm's Law states that the current through a conductor between two points is directly proportional to the potential difference (i.e. voltage) across these two points and inversely proportional to the resistance between them.

$$I = \frac{V}{R}$$

This 'Law' (not really a Law since there are materials that behave this way and some that do not) comes from German physicist Georg Ohm who wrote on it in 1827. His work involved the current through simple circuits in wires of varying length.

A very common analogy from hydraulics is drawn to illustrate Ohm's Law. Here water pressure, measured in Pascals (or PSI) is the analog of voltage since establishing water pressure difference between two points along a (horizontal) pipe causes water to flow. Water flow rate (liters per sec) is the analog of current (coulombs per sec) and flow restrictors, such as apertures placed in pipes placed between points (or other obstructions) where pressure is measured are the analog of resistors. Hence the rate of water flow through the aperture restrictor is proportional to the difference in water pressure across the restrictor. This is such a good analogy that flow and pressure variables can be calculated in fluid flow network by use of the hydraulic ohm analogy.

Our set of **Activities** here takes the basic circuit using batteries, resistors, wires and we measure the current and voltage of the circuits we create. In the **first activity** we explore Ohm's law for a single resistor and then test the idea for series and parallel circuits.

To help, note that in a **Series Circuit**, it is an electric circuit arranged so that charge flows through each component in turn. With resistors in series, all of them will act as a single resistor that is the sum of the resistors connected this way. This is known as the equivalent resistor. Each resistor has a drop in voltage, as it is using this much power of the supply. The current in a series circuit is the same for all components. Also, if one part goes out or becomes disconnected, then the circuit stops working.

In the case of the **Parallel Circuit**, it is an electric circuit arranged so that all components are connected to the same two potential points of the circuit so that each has its own branch. This means that if any branch becomes disconnected then the others will still operate. Each branch with resistors has the same voltage, yet a unique current depending on the amount of resistance in that branch line. If the components are all resistors in parallel, then the equivalent resistor for them is found as the inverse of the sum of the inverses of the resistances in parallel.

Purpose : To investigate the relation of voltage and current in a Resistor in a circuit of increasing measured voltage and the corresponding current for that circuit to graphically determine the resistance of the Resistor (acting in an Ohmeric fashion) to demonstrate Ohm's Law

Purpose : To measure the voltage and current to compute power use in a circuit of 2 resistors powered by batteries in a series circuit.

Purpose : To measure the voltage and current to compute power use in a circuit of 2 resistors powered by batteries in a parallel circuit.

NOTE : For any child, one must have permission and adult supervision when dealing with electrical equipment and devices. ***This Activity only involves DC circuits and there is no use of AC systems. Do NOT play with or plug anything into any wall sockets.***

NOTE : Only general descriptions of set up are given and it is assumed that one has some basic knowledge of resistors, circuits, multi-meter operation and use, series and parallel circuits, and the like. If in doubt, read up on it, seek reliable advice, and think through each step. Reread the directions as needed and look at the pictures provided.

Materials :

- Multi-Meter (measures DC Volts, Amps, Ohms),
- Alligator Clip-end Wires (as many as needed, 3 minimum),
- 4 AA Batteries & one 9-volt Battery,
- Battery Holders (1, 2, 4 AA batteries and 9-volt battery),
- Switch for Circuit (if wanted),
- Two 100 Ω Resistors,
- Graph Paper,
- Ruler,
- Slide Rule

Set-Up Procedure for the Activities :

1) For all cases, exercise safety and caution.
2) For all cases, be sure to know the voltages being used, the voltage and amperage ratings for the bulbs and resistors in use. Also estimate the current in a given circuit by using Ohm's Law ($V = I*R$) and **do not exceed 300 mA**. (check your multi-meter for its capacity level)
3) For all circuits, disconnect when the measurements are done.
4) In the case of the multi-meter be sure to know how to properly set up the meter to read voltage or current. Also be sure to start at the highest settings so that there is a smaller chance of damaging equipment when things are in operation.

Procedures :

1) **Activity 1 :**
2) Start with the single battery pack and
3) Attach alligator clips to the wires of the single battery pack .
4) Attach the other ends of the alligator clips to the resistor. Write down the value of the 100 Ω on the data table.
5) Now plug in the single battery to have the circuit operate.
6) Use the multi-meter and measure the Voltage (V). Note that this reading is taken across the resistor (that is in parallel to the circuit) across the resistor. Record this value.
7) Change the multi-meter and then Connect the multi-meter back into the circuit so that it is in series with the resistor to measure Current (I).
8) Before measuring Current, do a quick mental or slide rule calculation to estimate the current. Be sure that it will not exceed 300 mA.
9) Measure and record current (I).
10) Disconnect the battery pack and attach the next higher voltage version of it (that is go from the single battery pack to the double to the quadruple and finally to the 9V connector).
11) Redo all of the steps until you have completed the run of batteries up to the 9 volt battery where you measure both the voltage and the current for each set up.
12) Calculations to be done :
13) The first one we will do with the Slide Rule is to Calculate the Power for each of the circuits you have constructed and measured.
14) Next, Graph Voltage versus Current. Be sure to measure Voltage in Volts and Current in Amps. This means you have to recognize the readings on your multi-meter, such as mA meaning 1/1000 A, for example.
15) (Note : Science purists realize it should be Current on the y axis and Voltage on the x axis instead. If you wish do it this way, but realize after the slope, you need the inverse of it for comparison to the resistance of the Resistor! – For those who like the slide rule, in this case use the C1 scale for the slope and find the inverse on C or D below it)
16) Draw a best fit line for the graph and find the slope. Compare the slope to the resistance of the Resistor. Calculate percent error.

Activity 2 : Resistors in a Series and in a Parallel Circuit

Procedure :

1) **Series Circuit :**
2) Set up the Series circuit with the 2 resistors, the switch, and the battery pack with batteries. Connect the wires firmly so that the two are joined and one end from one is available as it the opposite end of the other resistor.
3) In the Series circuit there is only one path so the current should flow from the battery pack, through the switch when closed through each of the resistors in succession (check picture).
4) Note : For both the Series and especially the Parallel Circuit use no more than 6 V.
5) It is best to use two 100Ω Resistors.
6) Activate the circuit.
7) Measure the voltage across the battery connections and record this on the data table (V_B).
8) Now measure the voltage drop of each of the resistors (V_1, V_2) and record the values in the table.
9) Connect the multi-meter correctly so as to measure the current (I) for the series circuit and record this value. Since it is series, only one measurement is needed as all components in the series circuit have the same current.
10) **Parallel Circuit :**
11) Reconfigure the circuit so that the resistors are now in parallel to the battery pack. (see photo) This means that each of the resistors have their own branch.
12) The best way to do it is to use alligator clip wires so that each of the resistors is its own path.
13) When fully assembled with batteries, the switch (if desired), Activate the circuit.
14) First measure the voltage (V_B) across the battery pack and record this value
15) Check the branch voltages (V_1, V_2) for each of the resistors. Note these should ideally be the same. (Do not need to record)
16) For each branch measure the current (I_1, I_2) and record these values. Be sure on connecting the branch for the given current.
17) Calculation Steps :
18) For both the Series and Parallel Circuits, calculate Power.
19) Like the single Resistor, the Series and Parallel can have a range of voltages and measurements (From 1.5 V to 6 V) and the idea of determining the overall resistance of the circuit can be computed through graphing the Voltage on the y-axis and the Current on the x-axis.
20) Through the data points for each of the graphs (Series & Parallel) draw a best fit line and compute slope.
21) In the case of the slope for the Series Circuit, it should be equal to the sum of the Resistors in Series for that circuit.
22) In the case of the slope for the Parallel Circuit, it should be equal to the inverse of the sum of the Resistors in the Circuit.

Activity 1 Photo :

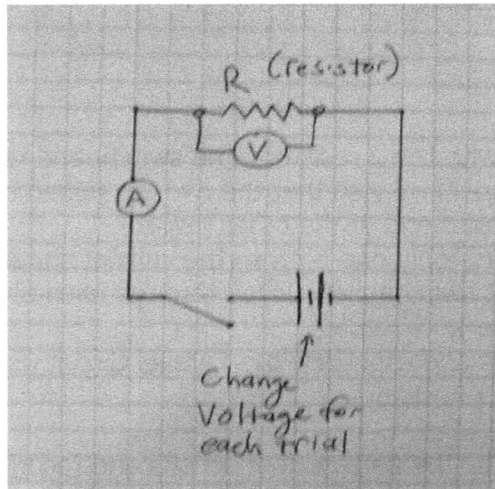

Series Circuit Photo :

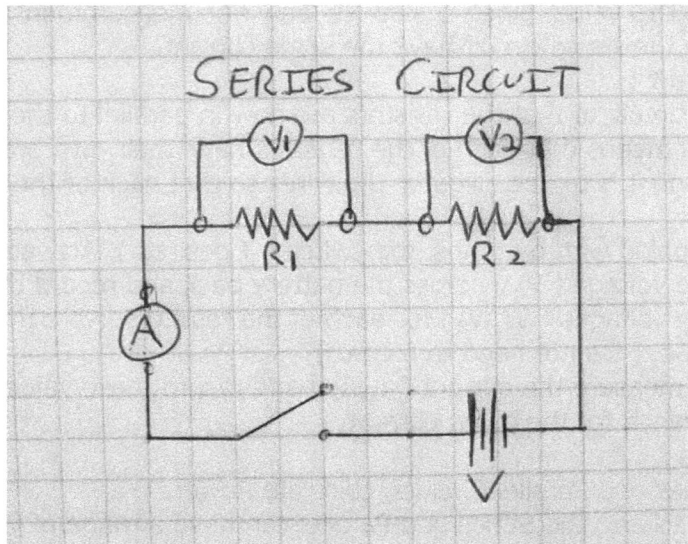

Parallel Circuit Photo :

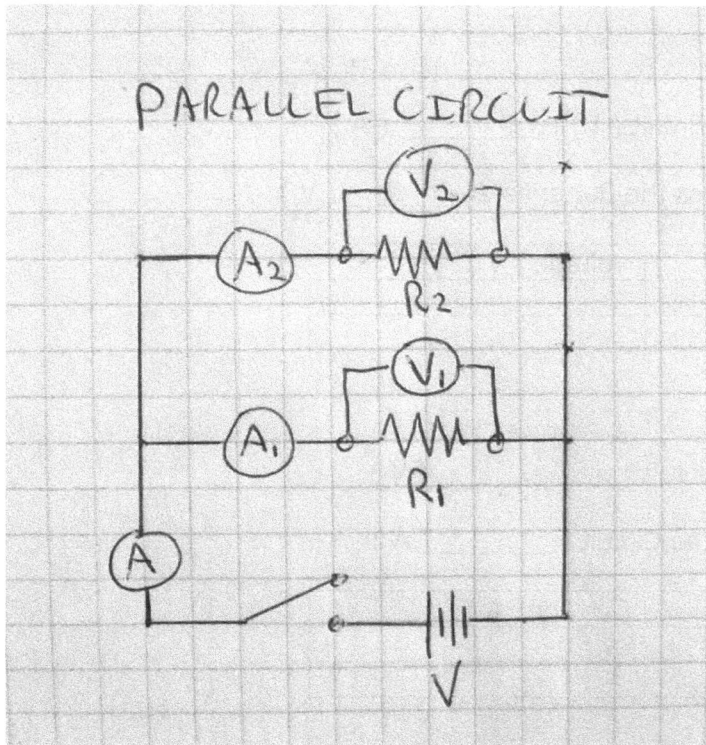

Data : Activity 1

Trial	Voltage (V)	Current (A)
1		
2		
3		
4		

Comparison Data :

Resistance of Resistor labeled : _____ (Ω)

Activity 2 : Series & Parallel Activity

Series Circuit :

Current for Circuit (I) : _____ (A)

Voltage across the Batteries : _____ (V)

Item	Voltage (V)
Resistor 1	
Resistor 2	

Parallel Circuit :

Voltage of Circuit : _____ (V)

Current for the Circuit : _____ (A)

Item	Current (A)
Resistor 1	
Resistor 2	

Calculations :

Be sure to use your Slide Rule!

Slope :

$$m = \frac{\Delta Y}{\Delta X}$$

$$\textbf{Resistance} = \frac{\Delta \textbf{Voltage}}{\Delta \textbf{Current}}$$

Ohm's Law :

$$V = I*R$$

Power :

$$P = V*I$$

$$\%E = \frac{[\text{Experimental Value-Accepted Value}]}{\text{Accepted Value}}*100\%$$

Formulae for Extension Activity :

Series Circuit :

$R_{eq} = R_1 + R_2$

Parallel Circuit :

$R_{eq} = \dfrac{R_1 * R_2}{R_1 + R_2}$

Conclusion :

What does the graph show as voltage is increased? How close is your slope to the value of the resistor being used?

Extension Activity Summary of idea and results:

The Series and Parallel Activity should have used **two resistors** (both **100 Ohm rating**).
first place them in **Series** and another in a second set of trials in **Parallel**.
For each circuit - Have a range of Voltages run from 1.5 V to 6 V and measure the Current in each case.

Record the Voltage and Current Readings as you did in Activity 1.
Go through all of the other steps noted above.

The key is to find what the slope of the Voltage vs. Current Graph finds for the Series Circuit of these Resistors.
You should find that it is the sum of these Resistors! This is because in a Series Circuit, the Equivalent Resistance is the sum of the Resistors in series with each other. So the slope will be 200 ohms ideally.

In the case of the Parallel Circuit case you should find that when using two equal resistors (say 100 ohm, which is the best choice here), the slope will come out to 50 ohms ideally.

This is because in a Parallel Circuit, the Equivalent Resistance is found to be equal to the inverse of the equivalent resistance equal to the sum of the inverses of the resistors in question.

Activity # 11
Electromagnet Graphing Explorations
Grade Level : High School
Math Level : Calculating

Electromagnet

One of the earliest items of fascination for a child is the magnet. Before long, one finds that they attract only certain types of matter (iron containing) and yet not others (paper, plastic, glass). Further investigation we uncover that there are two poles, North and South and the simple rules for them : Opposites attract and Likes repel. Further reading and learning and the magnet is not only an item on our refrigerators, they are found in many items : seals for doors, speakers, microphones, electric motors, electric generators, and so on.

Further we did not invent them, they are a natural part of nature. The Earth has a magnetic field generated in its current-carrying outer liquid metallic core, which extends far into space beyond the surface of the Earth and acts as a protective shield from harmful ionized particles from the Sun. Even the Sun and a number of the planets have magnetic fields, some like Jupiter's are much stronger than the Earth's field.

There are 3 basic categories of magnets : permanent, temporary, and electromagnet. All of them exhibit magnetic behavior due to moving electric currents (on a small or a large scale). The permanent ones have outer electrons in the metals that can move from one atom to the next in small loops called domains. This generates an electric current due to the electric field and perpendicular to it is a magnetic field. In the case of the electromagnet, the magnetic field is found and described by **Ampere's Law** which paraphrased states that the magnetic field strength around an electric current is directly proportional to the electric current in that wire or system.

First note that the electric current (field) and the magnetic field are at right angles to each other. A current carrying wire has a magnetic field around it. Use your right hand and wrap it around the wire and allow your thumb to point in the direction of the current. Your fingers wrapped around the wire are in the direction of the magnetic field. (This is classically called the Right Hand Rule)

Why then are the electric lines in our homes not acting as magnets? Simple there is a line going into the device and one coming out in the same wire. The magnetic field points in opposite directions and nearly cancels out.

The best way to make the electromagnet is not to use a straight wire, but instead to loop the wire. A coiled wire with current flowing will have a magnetic field where all the magnetic fields of the portions of the wire add together and generate a magnetic field. The more loops, the greater the strength of the magnetic field. Note that the electromagnet is only active when the electric current is flowing. This type of device is often called a solenoid.

The **use of the electromagnet** is extensive and found in large electromagnets to move metallic masses, cathode ray tubes or more commonly called conventional TV tubes where the electron beams are moved by electromagnets, galvanometers which are devices used to detect current / voltage by using a coil of wire in a magnetic field and the current in question passes through the coil to generate an opposing magnetic field to cause the deflection of a needle in the meter to render a reading (hence the basis of a voltmeter, ammeter, multi-meter), magnetically levitated trains due to electromagnets, electric transformers (where a current carrying-wire generates a magnetic field and this in turn generates an electric current of a different amount in another wire)[note this occurs in speakers too], and even in electric motors where current-carrying wires have magnetic fields which interact with the magnets in the motor to cause the rotor to spin (through the basic principles of magnetism – opposites attract and likes repel).

As can be noted by this last idea, electricity and magnetism are inter-related phenomena. One causes the other and vice versa. **This is the basis of the 2nd Industrial Revolution and the development of electricity, all of its appliances, tools, and applications in the modern world.**

Our activity involves the creation of an electromagnet from a current-carrying wire supplied by a power source (battery) and investigating the factors that may affect the overall strength of the electromagnet.

This sort of activity should inspire questions, though. If a current-carrying wire can generate a magnetic field, can a magnetic field induce a current in a wire? The answer is yes! This is the basis of **Faraday's Law**. Michael Faraday noticed that a wire when moving in a magnetic field will have a current generated or induced in it. It can be paraphrased thus :

In a place of a changing magnetic field there will be generated or induced an electric field. The strength of this electric field will be directly proportional to the rate of change of the varying magnetic field.

What this means is that the strength of the voltage will vary with the rate of the change of the magnetic field that the material experiencing the magnetic field will have created in it. A further analysis of this law of induction finds that the number of coils of wire will directly affect the induced electric field also known as emf (electromotive force) outcome as well. The larger the number of coils, the stronger the stronger the electric field.

This idea is the basis of the **electric motor** and **electric generator**. Each involves rotating wires and magnetic field. The motor converts electrical energy to mechanical energy while the generator converts mechanical energy into electrical energy. Each relies on the interaction of changing magnetic fields as caused by moving wires where one has current put into it (the motor) while the other has the current induced in it due to rotation (the generator).

Note : See Safety Notes below. This Activity creates a basic Electromagnet, which can cause batteries to heat up. This must be monitored carefully. It is only done with adult permission and supervision.

Note : It is imperative to be able to break or stop any circuit you create for this activity – i.e. a switch or the means to disconnect a wire – due to the fact that this type of circuit can heat batteries and cause damage to them. Leave no circuit unattended or active for too long a time – only short time uses.

Overall Purpose : To use the ampere's law to create a magnet from an electric current and test its strength by changing the variables that affect the formation of the magnet.

Purpose : To use an electromagnet and measure the strength of the magnetic field through changes in the number of windings of the wire making up the magnet.

Purpose : To use an electromagnet and measure the strength of the magnetic field through changes in the number of batteries being used in the magnet.

Materials :

- Iron Nails [as many as 3 different lengths are good here],
- Two D batteries & one 9V battery,
- Insulated Copper Wire (22 gauge type and 1-1.5 m length needed),
- Knife-switch Box (for circuit activation),
- Alligator Clips Wires (2),
- Wire Stripper,
- Battery Connection Boxes (single, double),
- Box of standard Paper Clips,
- Graph Paper,
- Ruler,
- slide rule

Procedure :

Activity I : The Number of Wire Windings vs Electromagnet Strength

1) Choose an iron nail will be used for this activity.
2) For this activity it is best to use 1 D size battery. Put the battery in the connection battery box and connect it to the knife switch with alligator clip wires on both ends. One will go to the knife-switch and then another will lead away. This one will attach to one end of the wire used to wrap the iron nail while the other end of that wire attaches to the alligator clip coming from the battery connection box.
3) Take the length of wire and wrap tightly the metal rod (nail) some 8 times and leave the other ends out. Record the number on the chart. Record the number of times wrapped (W).
4) Strip about 1" of the ends of the wire so that they can be connected to the alligator clips.
5) Connect the ends of the wire to the clips and then close the switch to activate the circuit. Have the dish of paper clips (all separated) and bring the electromagnet near them.
6) For each trial decide on a distance to be (0.5 cm) and go over the surface of the materials to pick up. Decide on the number of passes (1 or 2).
7) Pull the collection on the magnet off to the side over a cup or dish and deactivate the switch. Brush off the materials into the empty dish and then count the number of items picked up. Record this number (N).
8) Redo the wrapping process in step #3 now increasing the number of times (go from 8 to 12 to 15 to 20 to 25 to 30, perhaps up to 50 if possible). A good suggestion is a logical sequential increment such as 3 or 4 per trial.
9) For each trial record the results – (both the number of windings, W, and the value of paper clips picked up, N).
10) Graph the results where the Strength of the Electromagnet (as determined by the number of paper clips) is the y-axis while the Number of Wire Windings on the Nail is the x-axis.
11) From the graph, draw a best fit line and calculate the slope using the slide rule.

Activity II : The Amount of Voltage vs. Electromagnet Strength

1) In this activity you have chosen a base number of windings for the nail. Look at your data in Activity 1 to make this choice. Typically 15 to 20 is a good value.
2) Be sure to have the battery connection box attached to the knife switch and open. See Activity 1 for directions to the initial set up. You are creating a series circuit basically in each of these cases.
3) Set up the wire-wound rod and attach it to the power source. Activate the switch.
4) Pass the electromagnet over the materials to be picked up (paper clips or ball bearings). Be sure with each situation of data collection you maintain the same number of passes (predetermined as in activity 1) and the distance for each of the passes (predetermined as in activity 1).
5) Collect, count, and record this number (N).

6) Use the next size battery connection box (2 batteries). You will have to reattach the alligator clip from the single box to the 2 battery box.
7) Redo the work and test the strength of the electromagnet once again.
8) When completed with the D batteries, now use the 9 volt battery. For this one, the alligator clips can come from the wires themselves and attach to the 9 volt battery directly.
9) Record all the data,
10) Graph the data in the same manner as activity 1 (here the x-axis is the increase in voltages) and determine the slope of the best fit line with a slide rule.
11) Note :
12) Though the Slide Rule is a recommended tool, all of the calculations can be done with a graphing scientific calculator or the use of a spreadsheet program. In these calculations you have to generate a table of data, graph it, and then find the slope and/or equation of the best fit line for the data. Other formula calculations can be done with these tools as well.

Safety :

1. Adult Supervision and Safety are always important in activities.
2. Keep voltages low and Never use high voltage here as it can be dangerous and electrocute you.
3. It is important to activate the switch after making the loops and do not activate it without loops since the current in the wire will heat the battery which could lead to an explosion.
4. Monitor the temperature of the batteries. Do not leave them connected and do not have them on too long. They will heat up.
5. Activate the circuit to create the electromagnet for only short periods of time and monitor the battery so that it does not become too hot.
6. Never insert the wires into any other electrical outlet or devices.

Data :

Activity I : The effect of the number of wire windings on the strength of the electromagnet.

Voltage Used : _____
Number of Times Passed Over Paper Clips : _____
Distance of Each Pass Over Paper Clips : _____

Number of Wire Windings (W)	Number of Paper Clips Picked Up (N)

<u>Activity II</u> : The effect of the number of batteries on the strength of the
electromagnet.

Number of Wire Windings Used : _____
Number of Times Passed Over Paper Clips : _____
Distance of Each Pass Over Paper Clips : _____

Voltage Used (V)	Number of Paper Clips Picked Up (N)

Calculations :

Be sure to use a Slide Rule!

For all Activities, Graph the Data recorded where the y-axis is always the
Number of Paper Clips Picked Up (as it is the Dependent Variable), and
the Variable controlled (i.e. the Independent Variable) is the other one
on the x-axis – which will either the number of windings or the voltage used.
Draw a best fit line and determine the slope of the line in each case by using
a Slide Rule.

Slope = Rate of Change of Magnetic Field Strength
(note field strength is indicated by the amount of mass it
affects here in this activity)

$$\textbf{Slope} = \frac{\textbf{Number of Paper Clips Picked Up}}{\textbf{the Controlled Variable}}$$

Conclusion :

Examine the data and determine which, if any, of the variables has
an impact on the strength of the magnetic field. Can you write a
statement that generalizes the relations you may have uncovered. As the number
of windings increased, what happened to the number of paper clips that could be
picked up? How does the number of paper clips picked up relate to the inferred
strength of the magnetic field generated?

Activity #12
Creating and Measuring an Electric Generator
Grade Level : High School
Math Level : Challenging

This Activity is a follow-up to the Electromagnet Activity where more information in the prelude about such ideas and items can be found. Other Activities on electricity and magnetism have other information as well, such as Electrostatic Charge Activity and the Electric Motor Activity.

Here the basic idea is to use a small 35 mm film canister, some wire, strong magnet(s), and the like to create a changing magnetic field passing some wires wrapped about the canister due to a back-and-forth motion of one's hand holding the canister. The rate of change of the magnetic field's affect on the wires will be measured as electric current and voltage with a multimeter attached to the wires encircling the canister.

Activity I :

Purpose : To note changes in induced voltage and current in a loop of
wire due to the change in strength of the changing magnetic
field used to induce voltage and current.

Materials :

- Multimeter,
- Magnet Wire (22 gauge or larger)[100 to 200 cm total length],
- Wire leads with alligator clips,
- One 35 mm film canister,
- Set of small but strong magnets (3) (neodymium),
- Scissors,
- Duct Tape,
- Clear Tape,
- Cardboard (to be cut as rings),
- Measuring Tape,
- Clock with a second hand or Timer,
- Slide Rule

Notes : 1) This is NOT to be connected to any electrical outlet
2) you must have parental supervision and permission for all parts of the Activity 3) It is recommended to use 30 gauge wire (aka magnet wire)

Set Up : Create the Generator :

1) With the lid off the 35 mm canister, from the top and bottom measure about ½" so that there is a middle region on the canister that is about ½" wide. Mark these boundaries all the way around the canister with a permanent marker.
2) In the upper and lower region, use duct tape, cut to size to fit, and wrap several layers so that it becomes a boundary within which the wire will be wrapped and stay put.
3) (Note if not tall enough, the wire will spill out, so make it as tall as possible. If using too much tape, then use cardboard barriers cut to shape as needed where the barrier is a donut-shaped disk that encircles the canister and projects like Saturn rings out about ½").
4) Begin the wire wrap with 1-2 ft off of the canister and begin wrapping. The larger the number of coils, the better the generator. Keep the wrap as tight as possible and within the boundaries set up (if you need to wrap more duct tape or even use cardboard barriers).
5) When fully wrapped, have the other end of the wire extend about 1-2 ft as well. Tape the wrapped coil with some clear tape to hold it in place.
6) Scrape the enamel coating from the wire to create contact points.

Procedure :

1) In sequence into the created generator (see prior page) place one magnet at a time and seal with the lid to the 35 mm canister.
2) Connect the lead wires to a multimeter so as to read voltage first and then reconfigure so as to read current next. Measure in AC settings.
3) In both the voltage and current cases, you must shake the canister with the magnet in it back and forth in exactly the same motion. To help this process, use a clock with a second hand or a stopwatch so the periodic (back-and-forth) motion for each case (voltage and current readings) is the same. The rate can be whatever you choose.
4) Note : The faster the rate, the greater the readings will be for both. This in itself can be further studies done by you if you wish.
5) Record the reading values for your trials for both Voltage (V) and Current (I) on the table.
6) Calculate Power (P).
7) Each trial will be a different number of magnets (1, 2, 3, et al).
8) Be sure to have the same rate of motion for each trial so that comparisons can be made.
9) A bar graph of Power versus the Number of Magnets used in your generator can be made.

Data :

Number of Windings Used : _____

Trial (#Number of Magnets)	Voltage Reading (V)	Current Reading (I)

Calculation :

Be sure to use Your Slide Rule! :)

Power :

P = V* I

Conclusion :
The Electromagnetic generator should vary with number of magnets.
 Other experiments can be performed (such as rate of motion) as well
 as number of windings, etc.

Activity II : Alternative Generator Activity

Purpose : To measure the voltage induced in a given number of coils of wire by a changing magnetic field and the change of voltage induced by an increasing magnetic field.

Purpose : To measure the voltage induced in a given number of coils of wire by a given changing magnetic field and the change of voltage induced by an increasing speed of change for a given magnetic field.

Purpose : To measure the voltage induced by a changing magnetic field and the change in voltage induced by an increase in the number of coils of wire involved with a consistent magnetic field flux.

Materials :

- Multimeter (used for Voltage & Amp reading) (have alligator clip ends),
- Magnet Wire (22 gauge or higher is best – 26 or 30)[4 m total length],
- Paper Towel Tube,
- Plastic Crate that has holes on all sides,
- Set of small but strong ceramic magnets (at least 3, up to 5),
- Wire stripper,
- Scissors,
- Clear Tape & Masking Tape,
- Measuring Tape,
- Pillows or Cushions (Styrofoam, crushed newspaper, towels),
- Slide Rule

Note : Be sure to have parent permission and supervision for this activity. Pay particular attention to stripping the ends of the magnet wire. Also, as with the other electrical activities – these wires are not to be placed in any electrical outlet nor device for any reason.

Note : It is best to create a space for the experiments where the magnets have the least amount of probability of hitting a hard surface, hence plenty of cushioning with pillows, cushions, foam padding, crinkled up newspaper, and the like is suggested. The longer it takes the dropped magnets to slow down, the smaller the force affecting them (conservation of momentum concept).

Note : Be sure to run small tests on what works best for the configuration of the materials. Do several trial runs.

Set Up :

1. Note : It is best to use as strong a magnet set as possible, yet maintaining small size so that it fits into the paper towel tube. Also in the case of the sire, the higher the gauge the better.
2. Invert the crate.
3. Wrap enough wire around the paper towel tube so that it can be used for several of the experiments (such as 100 times or more).
4. For a given Activity, let the wire just remain as excess when not in use as a coil (for example, have 25 windings around the tube, but the remainder acts as a small pi e nearby.
5. Note : If this fails, perhaps due to too much resistance when measuring voltage and/or current, then have enough wire to cut the lengths needed for regular increments that you have chosen for your experiments (for example : 10, 25, 50, 100).
6. Cut the wire leaving some excess. Use the wire stripper to remove the protective coating on the wire.
7. Tape the wire with clear tape to the tube for each of the trials. Note that this will be removed in some of the cases, so it may need to be replaced regularly.
8. Attach the tube to the inverted crate with some masking tape so that it is vertical to the floor.
9. Place the inverted crate over your pile of cushions, pillows, and/or paper to act as a landing spot for the magnets.
10. Even before measuring the voltage or current, drop the magnets from up to 1m in height onto the cushioned area making sure it is going to be adequate for the task at hand.
11. Also test the passage of the magnets through the tube onto the cushions. A good idea is to use a second or a third tube as needed (note you may have to cut it to the needed height) to act as a guide for the falling magnets so as to keep them on course. The best idea is to use paper and roll it into a tube of needed height and tape it with clear tape to the tube with the wire on it.
12. Now test the system by attaching the multimeter leads to the ends of the wire and then letting the magnets fall. Set the meter to read AC millivolts.
13. It is best to start with a minimal test so as to determine the lower end of capability of the system. For example, use 10 wire windings and 1 magnet and drop it from a height of 10 cm above the tube.
14. If no response, change one of the variables (probably the windings are best) until a response is measured regularly.
15. Once the system is operational, proceed with the various Activities noted below.

Procedure :

1. For each of the Activities in this set use the basic set up of the device as noted above with the changes noted in the descriptions below.
2. Note : The higher the gauge of wire the better for the Activities.
3. Note : Be sure to know how to set up and measure either Voltage or Amperage on the Multimeter.
4. **Activity A : Varying Magnetic Field Strength**
5. Note : In this Activity, the overall configuration of the items do not change. The only change in this Activity is in the number of magnets used. You can only use as many magnets as the tube will allow through them. (If small this will be up to 3-4).
6. What is important is that you begin with the least number of magnets that generates both a millivoltage and a milliamperage. This may be as few as 1, but may be 2 or even 3. It is best to have the strongest magnets one can find for this set of Activities.
7. In this Activity, the things that are held constant are :
8. Number of Coils of Wire : (N) : 25
9. Height from which the magnets are dropped : (H) : 25 cm
10. The variable that changes is : The relative strength of the magnetic field (B)
11. To measure the given magnetic field, do the following :
12. With (a) given magnet(s), hold them at a constant distance above a small pile of paper clips and pass the magnets over the paper clips at a consistent speed one time.
13. Let the magnet(s) pick up a given number of paper clips and record this value (M1).
14. As another measure take the magnet set for a given trial and place it in the a somewhat spread out set of paper clips and slowly draw it out to see how many it can carry. Count and record this number (M2).
15. Do the paper clip pick up for each set of magnets to be used in this Activity (1, 2, then 3 magnets).
16. Whichever value shows a marked change (M1 or M2) will be the relative measure of the magnetic field strength for each of the trials in Activity 1.
17. Starting with 1 magnet, drop the magnet through the coil of wires around the tube and hooked up to the multimeter set on alternating (AC) millivolt settings at the prescribed distance. Be sure to have adequate cushioning so as not to damage the magnets.
18. Record the voltage reading (V_x).
19. For each trial (1 magnet, 2 magnets, 3 magnets) do this 3 times and record the voltage reading from each and then average this amount for a given set of magnets (V_{ave}).
20. Graph the Average Voltage (y-axis) versus the relative magnetic field strength (can be simply the number of magnets, or some measure of them such as M1 or M2) (x-axis).
21. Draw a best fit line through the points and determine slope of the graph.
22. Note : One can also measure the milliamperage (in AC too) as well for the same number of magnets. Be sure to do this 3 times for situation and average the results too.
23. If measuring amperage, you can calculate the power for a given trial as well. In this case, you can graph Power (P) on the y-axis and the number of magnets (N) on the x-axis.

24.

25. Activity B : Varying Number of Coils

26. Note : In this Activity, the same basic configuration is kept, only here these variables are held constant :

27. Height from which magnets are dropped : (H) : 25 cm

28. The number of magnets used : (N) : 2 (can use 1 or 3, if preferred)

29. The variable that changes is : The number of coils (N)

30. It is best to start with 10 coils and then progress in numbers of coils in subsequent trials (10, 20, 40, 80, 100, 150, 200, etc). The length of available wire, time to conduct the Activity, and the like affect this consideration. (The actual starting and ending point of coils does depend strongly on the strength of the magnets, the gauge of the wire – so experiment to find the right combination at first and then proceed accordingly). Each is a trial and is the vaiable 'C',

31. With each trial, like in the previous activity, do the exercise 3 times, taking voltage readings and then determine the average voltage for a given set of coils.

32. For each trial (10 coils, 20 coils, 40 coils, 100 coils, et al) do each 3 times and record the voltage reading from each and then average this amount for a given set of magnets (V_{ave}).

33. Graph the Average Voltage (y-axis) versus the number of coils (C) (x-axis).

34. Draw a best fit line through the points and determine slope of the graph.

35. Note : One can also measure the milliamperage (in AC too) as well for the same number of coils. Be sure to do this 3 times for situation and average the results too.

36. If measuring amperage, you can calculate the power for a given trial as well. In this case, you can graph Power (P) on the y-axis and the number of coils (C) on the x-axis.

37. Activity C : Varying Speed (Change) of Magnetic Field Strength

38. Note : In this Activity, the same basic configuration is kept, only here these variables are held constant :

39. The number of Coils of wire used : (C) : 25

40. The number of magnets used : (N) : 2 (can use 1 or 3, if preferred)

41. The variable that changes is : The height from which the magnets are dropped : (H) :

42. It is best to start with 10 cm and progress by 10s in subsequent trials. Note that you have to always consider the possibility of the magnets hitting and deflecting from the tube or the crate and then landing on a hard floor, hence being damaged. In this set of trials, you might want to consider additional cushions, padding, et al that surrounds the crate as well. In all you probably do not want a total height of more than 1 to 1.5 m above the tube.

43. With each trial, like in the previous activity, do the exercise 3 times, taking voltage readings and then determine the average voltage for a given set of coils.

44. For each trial (10 cm, 20 cm, et al) do each 3 times and record the voltage reading from each and then average this amount for a given set of magnets (V_{ave}).

45. Graph the Average Voltage (y-axis) versus the drop height (H) (x-axis).
46. Draw a best fit line through the points and determine slope of the graph.
47. Note : One can also measure the milliamperage (in AC too) as well for the same height. Be sure to do this 3 times for situation and average the results too.
48. If measuring amperage, you can calculate the power for a given trial as well. In this case, you can graph Power (P) on the y-axis and the height (H) on the x-axis.
49. From the Activities engaged in compare and contrast the slopes to see which (if any) had the greatest value (hence the greatest change) per unit increment.

Data :

Note : Tables only for Voltage are given, but one can readily change it to Amperage as needed for other measurements.

Activity A :

Number of Magnets	M1 (sweep over paper clips pick up)	M2 (immersion in paper clips pick up)
1		
2		
3		

Trial (Number of Magnets) (N)	V_1 : Voltage 1 (mV)	V_2 : Voltage 2 (mV)	V_3 : Voltage 3 (mV)	V_{ave} Average Voltage (mV)
1				
2				
3				

Activity B :

Trial (Number of Coils) (C)	V_1 : Voltage 1 (mV)	V_2 : Voltage 2 (mV)	V_3 : Voltage 3 (mV)	V_{ave} Average Voltage (mV)
10				
20				
40				

Activity C :

Trial (Drop Height of Magnets) (H) (cm)	V_1 : Voltage 1 (mV)	V_2 : Voltage 2 (mV)	V_3 : Voltage 3 (mV)	V_{ave} Average Voltage (mV)
10				
20				
40				
Etc				

The basic design of all the Activities can use this form below, only changing the Independent Variable Column as needed to that which is being considered.

Independent Variable Being Examined : _____
(Magnetic Strength, Speed of Magnetic Field Change, No. of Wire Coils)

Trial	Independent Variable	Voltage or Amperage Measured
1		
2		
3		

Note : copy this data table as many times as needed and for the number of trials needed in each case for the given Independent Variable in an Activity

Calculations :

Be sure to use your Slide Rule!

Rate of Change of Voltage with Respect to the Independent Variable :

$$\textbf{Slope} = \frac{\textbf{Δ Dependent Variable (Voltage)}}{\textbf{Δ Independent Variable}}$$

➤ The Independent Variables are : Magnetic Field Strength, Number of Coils, and Speed of the Changing Magnetic Field

Determining an Average :

$$M_{ave} = \frac{\Sigma m_x}{N}$$

N is the number of items in a set of data,
Σmx is the total of the values for the data set

Power :

P = V*I

Conclusion :

From your experiments, which variable(s) effect the greatest change in induced voltage from one trial to the next? In taking advantage of these increases of voltage, which would be the easiest to implement, lowest in cost to use?

Activity #13
Rate of Spin of a simple Electric Motor measure
Grade Level : High School
Math Level : Challenging

Electric Motors Activity

The Electric Motor is basically the reverse of the Electromagnetic Generator, Alternator, or Dynamo. In the case of the generator either a wire coil is spun between the poles of a magnet or vice versa so that an electric current is induced in the wire coil that is taken advantage of. The power source is typically steam, water, or wind which converts mechanical energy into electrical energy. Read more of the ideas of electromagnetic induction which is a part of the history of the electric motor as well in the Electromagnet Activity #11.

The **Electric Motor** is essentially the opposite of the generator. A power source, usually electricity causes the wire coil (set between two permanent magnets) has its current create a temporary magnetic field. The pieces of the motor are organized so that the current flow that creates this magnetic field is opposite the magnetic field of the permanent magnets. Of course, like poles in a magnet repel, so this causes a force to act on the wire coil that is set along an axis of rotation. This force acting along a distance from a central axis is a torque which sets the coil in rotational motion. Once half-way around the current is made to reverse direction by a split ring and this recreates the process again and maintains the rotation back around where the process starts again.

Hence the electric motor turns electrical energy into mechanical energy! Just by its simple description it is easy to see that energy must come from a source and is used to do something else. Since nothing is 100% efficient, there are losses (hence the motor gets hot). This helps to understand conservation of energy. (see The Story of Energy ch. V).

Electrical Motors have a wide array of uses and can range in size from the size of smaller than a fingernail and used in a watch to larger than a car and are used to propel ships, power industrial fans, and have ratings in the millions of watts. Most are handheld in size and power such things as fans, pumps, household appliances (washer, dryer, refrigerator), handheld power tools (drills, table saws, et al). They are one of the chief result of the 2nd Industrial Revolution which was centered on electricity and its development.

The ideas of electricity and magnetism date to speculations in the 1700s but were worked on in the 1800s and came into mathematical relationships and descriptions with Michael Faraday as early as 1821. Faraday did a public demonstration of a free-hanging wire in a pool of mercury through which a current was passed. In the pool of mercury was a permanent magnet. With the current flowing, the wire rotated about the magnet. This meant that the current-carrying wire had a magnetic field encircling it. He developed Faraday's Law. Basically it states that the induced voltage in a coil of wire is proportional to the product of the number of loops of wire and the rate at which the magnetic field passing through the loops is changing. More generally it states that an electric field is generated in any region of space in which there is a changing magnetic field with time.

The best way to think of it is this : A changing electric field induced a magnetic field and a changing magnetic field induces an electric field. This description connects to light even, since it is one of the many electromagnetic waves (radio, ultraviolet, x-rays, infrared to name a few) and the wave that it is.

By 1827 with experimentation Anyos Jednik developed a device he called a "lightning-magnetic self-rotor" which has the basic parts of the electric motor today, the stator, the rotor, and a commutator. However, it was not until 1832 when William Sturgeon invented an electric motor capable of turning machinery. Emily and Thomas Davenport patent a motor design in 1837 in America and it was made for commercial use. The early engines had limited success due to the lack of power availability. They were DC systems and at this time there were no significant battery systems in place.

In 1855 the first electric motor car by Anyos Jednik came into being. Not much else was done, however, for some time and the modern DC electric motor took full form in 1873 by Zenobe Gramme and it was used in industry. More developments followed, such as 1888 when Nikola Tesla created the first practical AC electric motor.

With further testing, experimenting, and new ideas and understanding of electrical and magnetic forces (Maxwell's equations in the 1880s) plus the practical application of ideas to design led to stronger, more efficient engines. With this, industry could literally go into high gear since now it had a workhorse that had high reliability, power, and was efficient. It is the electric motor that changed the workplace to reduce the number of needed people, animals, and to replace older hydraulic pressure systems to run and operate all levels of machinery. The home life changed drastically in terms of ease and convenience. Even today the electric motors used in homes and industry accounts for more than half of all the electricity being used. Even today in the 21st century the use of the electric motor is once again finding its way into discussions involving motorized vehicles for transportation.

Our goal is to build and test a simple DC electric motor only using battery wire, paper clips, a magnet, and batteries.

Notes on Safety : As with any electrical device or situation, exercise caution. Be sure to have adult permission and supervision. Be careful of sharp objects (such as with wires). Do not place wires in other electrical devices, outlets, and the like. Be sure to wear goggles.

Purpose : To measure the spin rate variance of a constructed electric motor due to the amount of applied voltage to the motor.

Alternate Purpose :

Purpose : To measure the spin rate variance of a constructed electric motor due to the number of windings of coil wire in the motor.

Materials :

- Magnet Wire (24 gauge or higher) (total 6 m in length),
- Ceramic Magnets (3 of them each approx. 3 cm^2 by 0.5 cm thick),
- Block of Styrofoam (approx. 5 cm^2 by 2 cm thick),
- Battery Holder Packs (1 AA, 2 AA, 4 AA),
- Batteries : four AA and one 9-Volt,
- Wires with Alligator Clips,
- 2 Large Paper Clips,
- Sandpaper (safest method) OR Craft Exacto Blade,
- Laser Photo Tachometer (to determine spin rate),
- Goggles,
- Slide Rule

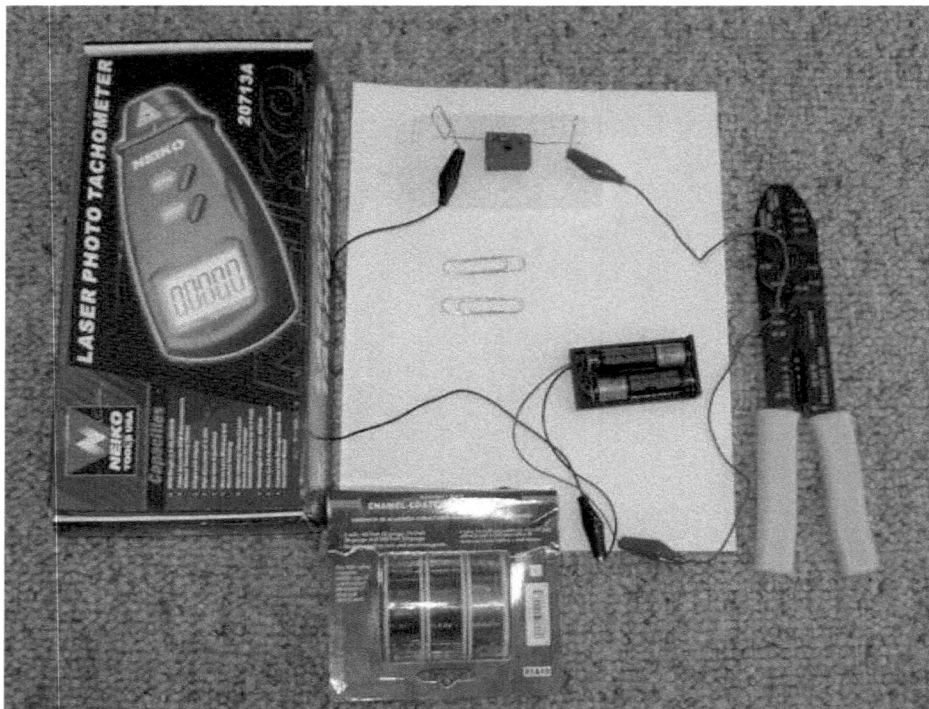

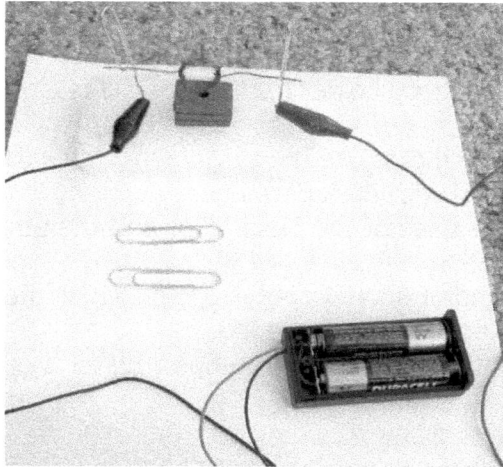

Set Up of Electric Motor : (Have parents do this)

1) Cut a long enough piece of magnet wire so that it can wrap around a broom handle or C battery about 10-20 times and have about 3 inches either side of this at most. (Length is around 36-48 inches total).

2) When the coil is created wrap the wire back around the coil so as to secure it and the the remainder of the wire project away from the edge of coil in opposite directions.

3) Be sure to refer to the pictures to help in the descriptions!

4) With the coil facing you flat face (like a coin) and the wires projecting at the sides place the projecting wire piece from each side in turn and use either sand paper or a sharp blade and scrape off the enamel coating on One Side of the Wire.

5) Do this to the other projecting wire and scrape The Same Side so that the top is scraped for both while the bottom is coated.

6) Into the foam block stick two large paper clips after straightening the outer coil of the outer loop. This creates two loops standing up.

7) Between the paper clips place the ceramic magnets.

8) Through the loops of the paper clips place the coiled loop of wire so that the projections pass through the paper clips and rests there.

9) For this and the procedure it is good to wear goggles.

155

Procedure :

1) Follow all of the assembly directions for the engine above.
2) Place the batteries in the battery holder (start with the least and work your way up in each of the trials.
3) Over the coil of wire on the motor use the tachometer reflective coil tape (be sure that the underside is coated with dark permanent marker)
4) Attach alligator clips to the base of the paper clips and the other ends to the battery pack.
5) With some agitation the motor should operate. Recheck all the work of the engine and test it before the reflective tape if needed.
6) Use the tachometer to obtain readings of rotation rate for a given voltage. Record these results.
7) With each trial try another voltage.
8) Graph the data and calculate slope of the best fit line of Spin Rate vs. Voltage for the Motor.
9) Note :
10) Though the Slide Rule is a recommended tool, all of the calculations can be done with a graphing scientific calculator or the use of a spreadsheet program. In these calculations you have to generate a table of data, graph it, and then find the slope and/or equation of the best fit line for the data. Other formula calculations can be done with these tools as well.
11) Alternative Trial
12) With more battery wire create another loop and run the tests again but here see how the effect of the number of windings affects the outcome.
13) Note here the voltage is kept constant, since the variable under consideration is the number of windings.
14) Perform the same basic calculations only here it is spin rate vs number of windings.

Data :

Activity I : Spin Rate with Varying Voltage

Number of Windings : _____ (20 recommended)

Trial	Voltage	Spin Rate
1	1.5 V	
2	3.0 V	
3	6.0 V	
4	9.0 V	

Alternate Activity :
Activity II : Spin Rate with Varying Number of Windings

Chosen Voltage : _____ (3.0 V recommended)

Trial	# Windings	Spin Rate
1	10	
2	20	
3	40	
4	60	

Calculations :

Be sure to use your Slide Rule!

Graph the Measured Data with the Independent Variable (the number of windings or voltage) on the x-axis and the Dependent Variable (the motor's spin rate) on the y-axis.
Draw a Best Fit Line.

Slope $= \frac{\Delta Y}{\Delta X}$

Slope of Line $= \frac{\Delta \text{Spin Rate}}{\Delta \text{Independent Variable}}$

Conclusion :

What did the results show? Do you think there are upper limits to this?
Why or why not?

The power and use of the electric motor cannot be underestimated and can be a great source of ideas, inspiration, and other lab activities. There are different materials and slightly different means of arrangement to achieve the same ends and even kits to build them.

Activity #14
Simple Magnetic Field Strength determination
Grade Level : High School
Math Level : Calculating

The Basic Magnetic Field Strength Investigation Activity

Magnetism is related to Electricity, so read the introductions to Electric Charge Activity (Activity #3) and the Electromagnet (Activity #11) and Electric Motor Activities (Activity #13) for more information, in general, on the connection of Electricity to Magnetism.

It essentially turns out that a moving current has an associated magnetic field. (Much like a changing magnetic field generates a current). Though we are using a permanent magnet in this Activity, the moving current is the aligned outer electrons in the material which result in the magnetic field for the magnet.

Like other forces (gravitational and electric), the magnetic force on a large enough scale should exhibit an inverse-square law characteristic. That is to say, as distance from the magnet, the force should decrease as the inverse-square of the distance. For more on the idea of inverse-square laws, see the Inverse Square-Law of Light Activity (Activity #20)(where light illumination from a point source acts in the same manner).

In this Activity we explore Magnetic Field strength in a rather simple manner. We use a Magnet separated from a piece of metal it can affect by some poster board paper barrier. The piece of metal is a part of a Tension Scale so that we can pull the scale and measure the amount of force that the magnet is exerting to hold the metal in place. We use different thicknesses of poster board paper, measure, and graph our results.

In the Activity, be sure to have parental permission and supervision. Be sure to wear goggles. Act carefully in a slow and deliberate manner for safety-sake as well as taking measurements.

Purpose : To measure the pull (force) of a magnet on a magnetic material at varying distances so as to uncover the relation of distance and field intensity through graphing and analysis.

Materials :

- 1-3 small ceramic or neodymium (strong) Magnets,
- 100g Tension Scale (depends on magnet strength),
- Poster Board,
- Scissors,
- Caliper or Ruler,
- Masking Tape,
- Graph Paper,
- Goggles,
- Slide Rule

Photos :

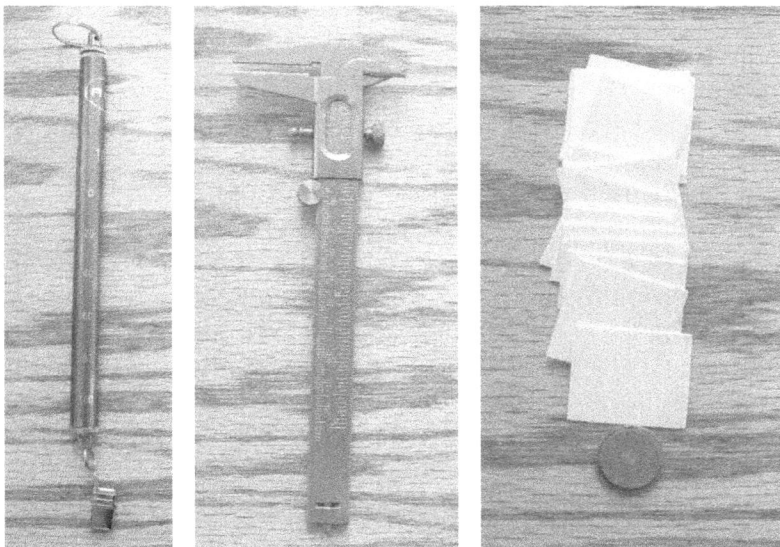

Procedure :

1. In the Set Up do the following :
2. Be sure to have a strong magnet that can effect items, like paper clips. through a number of sheets of paper. This is because we are using poster board (or something thicker than regular paper) in this activity.
3. Next test the Tension Scale and whether or not the hook or clip at the end is attracted to a magnet. If not, attach securely a paper clip.
4. Before the next step, and any step involving the magnet being attached to the metal hook/clip of the scale (even through the poster board, which is the investigation in this activity) you should wear goggles in case the hook/clip should spring back too quckly and/or possibly detach. Knowing this can happen, be aware and work safely. Also examine the tension scale – it is recommended that the hook/clip is securely attacned. Have a parent supervise the activity and inspect the tension scale.
5. Test the Tension Scale through several layers of paper (or even a piece of cardboard) with regards to the magnet attracting it and holding it.
6. In testing the Tension Scale, now slowly pull the scale so that it extends while attached to the magnet. At some point it should release. If you can fully extend the scale, then the thickness is not enough for the barrier between the magnet and the scale hook/clip or the scale needs to be of a higher order (1000 g instead of 100 g)
7. The final step in the set up is to cut small squares out of the poster board that are a bit larger than the magnet being used (see photo). The number will depend on the strength of the magnet – a good estimate is 25.
8. With the tests done, onto the Activity :
9. Zero out the Tension Scale.
10. It is recommended during and between given trials when it can be checked it is a good idea to check this and reset it as needed.
11. In each trial you will do the following :
12. Have chosen a given number of poster board barriers (referred to as barrier pieces in the data table) that will be between the magnet and the tension scale hook/clip.
13. It is best to start with the maximum and then diminish this number (as will be seen in the continuation of measurements) (for example 25). Record this number (N).
14. Stack the number of chosen pieces atop the magnet that is on the table top and tape them down. The tape will look like a bridge from the side and can easily be removed (as it will be in subsequent trials). Note that the top piece will probably be stuck to come along for each of the trials, but that is okay.
15. It is best to have this next to the table's edge, since you now need to use a ruler or a caliper to measure the thickness of the barrier and record this value (T) in centimeters.
16. Now bring the Tension Scale hook/clip down to the top of the stack and let it magnetically attach. Once secure slowly draw it away, all the while watching the reading on the Scale.
17. At some point the hook/clip will detach. Be sure to keep an eye on where the scale reading had gone to (M). Also be sure to watch your eyes by wearing goggles! Keep enough distance for safety.
18. Do the prior two steps 3 times, each time remembering and recording the tension scale reading. These values will be averages for a given trial (M_{ave}).

19. In the next and subsequent trials, remove the tape and remove some of the poster board barrier pieces to decrease thickness, and re-tape the new stack.
20. It is logical to have a pattern to the number of pieces, for example if you started with 25, with each trial remove 5 of the pieces, so that the next trial is 20, then 15, and down to 5 or zero if you like.
21. Recognize that with zero pieces, you will more than likely need a scale that is of a higher caliber, since the force is quite large at this point. If you do not have one, simply note that it exceeds your measurement capabilities and goes off the scale.
22. Record the new values of pieces (N) and thickness (T) along with checking the calibration of the Tension Scale.
23. Repeat the process of measuring the amount of tension force (M) and averaging the reading for a given trial (M_{ave}).
24. Once all trials are done, put materials away.
25. Calculations
26. Though we have not technically measured force (unless your scale is marked as such), we can either use the readings as they are or first convert them to genuine force measures (F) from the M_{ave} values. (Use Force formula and conversions as needed).
27. Graph Tension Force (either M_{ave} or the newly computed F values – hereafter simply called F) on the y-axis vs. the thickness (T) (which can be in centimeters or meters) on the x-axis.
28. In the case of this graph it should not be straight and in fact, if ideal will have a slope of an inverse-square (see Inverse Square Law Activity #41 for further illustration) or proportional to $\frac{1}{T^2}$.
29. One of the first ways to see how close this relation is to an inverse-square is to graph calibrated Tension Force values (F) by dividing all of the table values by the largest value in the table (this means the largest one will have a value of 1, all others will be less).
30. Using the data for Thickness from the same data points (the one with the largest Tension Force value, use its distance value as the calibration value and divide all Thickness data points by this value, (hence the smallest value should be 1 and all the rest should be numbers greater than 1).
31. Now take the inverse-square of the calibrated Thickness ($\frac{1}{T^2}$).
32. Note that the above directions indicate that you must reformulate your table where the y-value column is the calibrated Tension Force value which is determined by taking a given Tension Force value and dividing by the largest value in the table. Do this for each of the values.
33. In the case of the Thickness values, take the inverse-square of each of the calibrated Thickness values.
34. Note that all of these actions can be done on the slide rule (as should all of the calculations). Here one can use the C1 and then the A scales after finding the T value first on the D scale. Note that the C1 is very useful for inverses, and as for the divisions, simply employ the C & D scales.
35. On this graph we can draw a best fit line and determine slope, which ideally should be 1.
36. We now want to find out if it is indeed an inverse square, so now take our original data table and convert all values in it (F and T) into log values of each using the slide rule's L scale.

37. You will now have a table of Log(F) and Log(T) values.
38. Which you can now graph as well on a new graph of Log(F) on the y-axis and Log(T) on the x-axis.
39. Draw a best fit line through this graph and determine slope – it is ideally linear and should have a slope of -2.
40. You can do this same set of steps for other magnets (larger or more) and see if there is a difference.

Data :

Trial #	No. of barrier pieces : N	Thickness of barrier : T (cm)	Scale Reading M_1 (g)	Scale Reading M_2 (g)	Scale Reading M_3 (g)	Average Scale Reading : M_{ave} (g)	Force measure of scale : F (N)
1							
2							

Calculations :

Be sure to use your Slide Rule!

Average :

$$X_{ave} = \frac{\Sigma x}{n}$$

(x's are the values measured and 'n' is the number of times)

Slope :

$$m = \frac{\Delta y}{\Delta x}$$

$\Delta x = x_2 - x_1$

Force from Mass Measurements on Tension Scale :

$F = m*g$

(Note : 'm' here will be our M_{ave})

162

Conversions & Constants :

$1 \text{ m} = 100 \text{ cm}.$

$1 \text{ kg} = 1000 \text{ g}.$

$1 \text{ N} = 1 \frac{kg*m}{s^2}$

Acceleration due to Gravity : $g = 9.8 \text{ m/s}^2$

Conclusion :

What do your results show – that is, as distance increases, what happens to the magnetic force intensity ? Does it decrease in a linear manner, an inverse manner, an inverse-square manner? Will it matter if you change the type of barrier between the magnet and the scale? How did the force change when there were more or less magnets, different magnets? If you did the Inverse-square law activity, How does the graph of Force and Distance compare to the Light Intensity and Distance graph in the Inverse-square law of Light Activity?

Activity #15
Home-Made Galvanometer & Voltage Measurement Activity
Grade Level : High School
Math Level : Challenging

Galvanometers were devices that were developed when it was noted that the needle of a magnetic compass was deflected by current in a wire that was wrapped around the compass. This phenomena was reported by Hans Oersted in 1820. A galvanometer is one type of ammeter (a device used for detecting and measuring electric current). In common instrument a needle is attached to a moving coil so that when a current passes through the wire wrappings a magnetic field is generated and this is in opposition to a magnet in the coil so that a deflection of the needle through movement of the coil occurs. This comes from the fact that a current in a wire generates a magnetic field around it.

Noting the response of a magnetic compass needle now gave to scientists the means to measure currents and therefore compare their values and then be able to conduct experiments to see what effects the current of a given circuit or electrical system. The name galvanometer comes to us from Luigi Galvani who found as early as 1771 that an electric current could make a frog's leg twitch or move. These devices are also called 'multipliers' since the earliest ones had the magnetic field generated by numerous wire turns so as to increase the strength of the magnetic field generated. The first noted creation and use of this idea was by Johann Schweigger at the University of Halle in 1820. These earliest ones are called 'tangent galvanometers'. Like many other scientists involved in electricity, even Andre-Maria Ampere added to the idea of the device.

Not all of the galvanometers have the same configuration. A later type, called the 'astatic' version uses opposing magnets and does not rely on the Earth's magnetic field for operation. Probably the most sensitive of these types used a mirror, hence the name mirror galvanometer, where small magnets are attached to a wire also attached to a mirror which in turn would deflect the path of a beam of light so as to greatly magnify the deflection of the system due to very small currents. This was made by William Thomson (Lord Kelvin).

For those interested here is the magnetic field strength (B) of the system in the wire (note : in our Activity, we are not looking to measure this, we are merely using the device to detect and measure voltage, but the following mathematical information is provided to show the basis of the tangent galvanometer) :

$$B = \frac{u_0 * n * I}{2 * r}$$

Where u_0 is a constant, n is the number of turns in the coil, I is the current in amperes, and r is the radius of the coil

The expected angle of deflection, Θ, for a tangent galvanometer comes from this expression where B_H is the Earth's Magnetic Field :

$$\Theta = \tan^{-1}\left(\frac{B}{B_H}\right)$$

From the tangent law, $B = B_H * \tan\Theta$

From this we can derive :

$$I = \left(\frac{u_0 * n * I}{2 * r}\right) * \tan\Theta$$

$$I = K * \tan\Theta$$

K is the Reduction Factor of the tangent galvanometer

Galvanometers through time were built with more precise technology and have an array of applications. They are the first tools to find that there is electrical activity in the body, particularly the heart and brain. They also have been used to detect singles from long submarine cables.

What can be said of the galvanometer is the following :

1) In general, as current is increased, there is a stronger magnetic field generated, hence a greater deflection of the needle in use in the device
2) It can be used for voltage readings as well and follows the same general rule as above, where an increase of voltage will increase the angle of deflection of the needle of the system in a proportional manner. –

Note : In the case of the device now being a voltmeter (instead of an ammeter) this must happen : The galvanometer now has a Resistor of sufficient resistance so as to limit the deflection of the coil in the galvanometer.

The galvanometer is the basis of the multimeter today, both the analog and the digital models. In the analog it is clearly noted in the needle that deflects across the face of the gauge. The digital uses a converter to now display results on a screen as opposed to being read from the scale.

The multimeter today is used for current, voltage, resistance, and capacitance alike. The device and its operation does depend on how it is connected to the system or circuit in question as to its operation. This also includes the range over which it can operate for a given situation.

Our Activity has us develop a very simple tangent galvanometer actually using magnet wire and a compass along with a resistor in series so that it acts like a voltmeter. With the resistor in place, it will stop full scale deflection of the compass needle and we should find that with increasing voltages, there is an increasing angle of deflection of the compass needle. We can then calibrate our device from graphing measured data and then use it to estimate the voltage of other batteries that we may have on hand but are uncertain as to their values.

Note a couple of things : first that too small a current (from small voltages) and too large of a current (from high voltages) takes away from the precision of the tool. Second, be patient and willing to experiment with various resistors. You may even find that you have to connect a few resistors in series with each other so as to have the desired results. Third, be sure to have parental permission and supervision for such activities. Fourth, do not connect the wires and items to any AC system nor any high-powered electrical system of any sort. This device is merely used to examine common everyday batteries.

Purpose : To construct a basic galvanometer and use it to determine the Voltage from the current flow in a coil of wire that creates a magnetic field that conflicts with the Earth's magnetic field as noted on a compass causing in a direct current circuit and calibrate the instrument with a multimeter.

Materials :

- Multimeter,
- Compass (the larger and better the quality, the better the results),
- Magnet Wire (20 through to 26 gauge – pick one),
- Alligator Clip Wire Set,
- Battery pack for 1, 2, and 4 AA batteries for connection,
- Four 1.5 V AA batteries,
- One 9 V battery,
- Note : You will have to have other batteries, C – D – AA – AAA – other 9 V batteries to test once the 'device' is created and calibrated,
- Resistor (chose from range : 100 Ω or less),
- Protractor (if needed – best to have compass with degree measures),
- Regular Tape,
- Slide Rule

Note : Be sure to have parental permission and supervision in this Activity. Do not connect any electrical device or use wall outlets in any manner in this activity. It is only about using regular batteries. Also of importance – the circuits set up should be disconnected as soon as possible. Do not have them on for long time durations – only a handful of seconds to take measurements and that is all.

Note : The Protractor is only needed if the Compass does not have an angular measurement on it – this is one of the key measurements that is taken in this Activity.

Note : Notice the magnet wire and the resistors – one does not need each variety of all of these – it is best to choose no more than two and test them. As used in my example photos below, I used 20 gauge magnet wire and 100 ohm resistor. I do note that there was only a small angular movement of the compass at two batteries (even smaller with one battery), but with more batteries, there is greater deflection.

Set Up Procedure :

- Note A : Follow the steps that follow, but realize that it will all depend on your equipment, the testing, and adjustments you make to your galvanometer's design that will determine its success.
- Note B : Essentially this is a trial-and-error process where you may need to change things as needed. For example, you may have to have a different number of magnet wire wraps around the compass used, use a different compass, use a different set of resistors, and the like to achieve the needed results.
- Note C : For the set-up use a given and known battery set. Brand new and/or fresh is best for this. In the later tests, once these are used to create and calibrate the system, then examine other batteries to see how effective your device is at measuring the voltage of an unknown battery that you are testing (Note : Use only standard batteries, such as AA, AAA, C, D, 9 V, and with permission and supervision, of course).
- Note #1 : It is best to have a good quality compass that is approximately palm-sized and has degree measurements either on the needle part or the dial. Price ranges are usually over $10 and should be less than $70 – it really depends on your interest and uses for this later.
- Note #2 : The gauge of the magnet wire will affect the size of the resistor employed. You should probably buy at least two different gauges to find the one that works best.
- Note #3 : Cut a length of magnet wire that will wrap around the compass and leave about 3 in. from each end. The number of wraps is up to you, but keep it small (less than 12). – Be sure to strip the magnet wire at the ends – this may require sandpaper or a knife. Have adult permission and supervision and let them strip the wire ends.
- Note #4: Run a quick test by simply wrapping the wire and holding the wire ends to a battery at each of the ends of the battery to make a complete circuit. Align the compass and the wire wrap in the same direction. To demonstrate that there is a current in the wire and that it is generating a sufficient magnetic field that affects the compass. – It should have a large deflection of the compass needle. - Disconnect the circuit.
- Recognize that if the attempts keep failing it may be best to employ the multimeter and test the circuit's voltage and the batteries voltage values.
- Note #5 : You can tape the wrapped wires to keep them in a bundle.
- At this point connect the alligator clip wires to the ends of the magnetic wire.

- Now you can now test having a given resistor (can be as low as 10 ohms and as much as 100 ohms) connected to one of the alligator clip wires. Note : Use an alligator clip wire from the magnet wire to the resistor and then an alligator clip wire to the battery pack. Along the other path have one alligator clip wire attaching the battery pack to the magnet wire (see photos of set up below).
- To test be sure to realign the compass and the magnetic wire wrap.
- Now test by connecting the resistor free end to one end of the battery and the other alligator clip wire to the other end of the battery and note any movement of the compass.
- Try increasing the voltage on your preliminary system. You should find in subsequent needle deflections that it increases with increasing voltage.
- Note #6 : If the deflections are too great, increase the amount of the resistor in your circuit. Test until you find the adequate amount of resistance that allows for a small but measurable deflection of the compass needle.
- With all tests completed, now use the system in the Procedure to take measurements.
- Note #7 : Be sure to never leave the circuit active for too long a time. This can cause overheating. All tests should be of a short time duration and then disconnect the circuit. Also, you might have to have the circuit be hooked in the opposite direction so as to take away any residual magnetic effects affecting the compass. Be sure that the compass returns to its original settings after each test.

Photos of Set Up for Galvanometer :

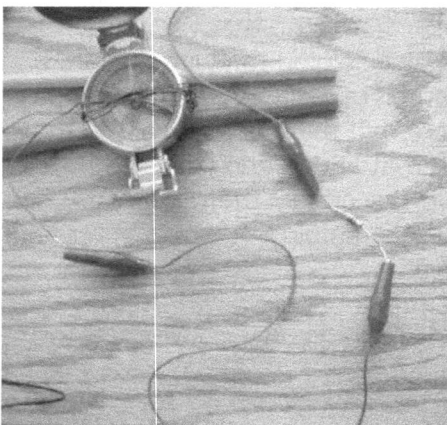

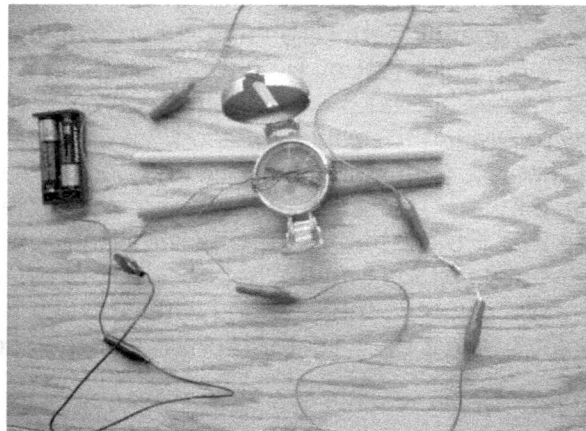

Procedure :

1. First start with the Set-up Procedure to create and test your Voltmeter (galvanometer).
2. The process of the set-up is not only done to create a working galvanometer, but also to record the necessary data for its calibration, which is the main goal of this Activity.
3. With a functioning home-made galvanometer / voltmeter ready to go :
4. First align the wire wrap on the compass and the compass needle. At this point there are No batteries in place.
5. Be sure to note from where you will measure the angle and call the initial line zero $0°$. Be sure to read the device in the same manner each time (for example, do not look straight on at first and then at an angle to the line of reference – be consistent).
6. Before attaching the battery system to the alligator clip wires, first measure the battery system (whether it's one, two, or more batteries) with the multimeter set at the proper setting of DC voltage (and proper range) to have a figure to calibrate with. (multimeter V)
7. Now attach the battery system to the alligator clips so that the current through the circuit (including the magnet wire and resistor) will be complete and cause a magnetic field to interfere with the compasses response to Earth's magnetic field and cause a deflection of some small angle.
8. Measure and record the angle of deflection (Angle $°$).
9. Repeat this process for each set of calibrating battery systems that you wish to use – 3 is enough, but you can do more if necessary.
10. Graph on a Cartesian coordinate system the results of your findings where the multimeter voltage (V) is the independent variable on the x-axis and the angular displacement ($°$) of the compass is the dependent variable on the y-axis.
11. Draw a best fit line through the points.
12. Determine the slope of the points.
13. Now have a series of unknown batteries that were not used for calibration to be tested for voltage by your voltmeter / galvanometer.
14. With each battery, place it in the system and measure the angle of deflection – record these results.
15. Use the formula for Predicted Voltage (V_p) to determine what your voltmeter / galvanometer is measuring as a result.
16. Now measure the test batteries to compare to your predicted results.

Data :

Multimeter Readings :

Trial	Battery System	Voltage Measurement from Multimeter (V)
1	One 1.5 V battery	
2	Two 1.5 V batteries in series	
3	Four 1.5 V batteries in series	
4	One 9 V battery	
N		

Compass Wrapped with Wire and Connected to calibrate instrument :

Trial	Battery System	Angular Movement Measurement of Compass (°)
1	One 1.5 V battery	
2	Two 1.5 V batteries in series	
3	Four 1.5 V batteries in series	
4	One 9 V battery	
N		

Tests on Unknown battery system Voltages (use other batteries) :

Trial	Battery System	Angular Movement Measurement of Compass (°)	Voltage Measurement from Multimeter (V)
1	One 1.5 V battery		
2	Two 1.5 V batteries in series		
3	Four 1.5 V batteries in series		
4	One 9 V battery		
N	Note any and all can be different than what is noted above in the list – these are the unknown ones		

Calculations :

Be sure to use your Slide Rule!

Slope for calibration :

$$m = \frac{\Delta Y}{\Delta X} = \frac{\text{Angular Movement of Compass}}{\text{Voltage of Battery System}}$$

Predicted Voltage : V_p

$$V_p = \frac{\text{Angular Movement of Compass for Test Batteries}}{\text{Slope from calibration}}$$

Conclusion :

How did your Galvanometer function? What sort of relation did you find with regards to changing the resistance of the circuit and the measurement of current? (What sort of angle of deflection occurred in each set of measurements as compared to the other ones)?

Activity #16
Home-Made Pendulum Magnetometer &
Magnetic Field Measurement Activity
Grade Level : High School
Math Level : Challenging

Magnetism is related to Electricity, so read the introductions to Electric Charge Activity and the Electromagnet and Electric Motor Activities for more information on the connection of Electricity to Magnetism. Both are described in the summary of Laws known as Maxwell's Equations. Both Electricity and Magnetism are very strong forces.

It essentially turns out that a moving current has an associated magnetic field. (Much like a changing magnetic field generates a current). Though we are using a permanent magnet in this Activity, the moving current is the aligned outer electrons in the material which result in the magnetic field for the magnet.

As with all magnets, they have a North Pole and South Pole. As we are often called upon to recall : Opposite Poles Attract one another while the Same Poles Repel one another.

Like other forces (gravitational and electric), the magnetic force on a large enough scale should exhibit an inverse-square law characteristic. That is to say, as distance from the magnet, the force should decrease as the inverse-square of the distance. For more on the idea of inverse-square laws, see the Activity, the Inverse Square-Law of Light #20 (where light illumination from a point source acts in the same manner).

In this Activity we explore Magnetic Field strength in a rather simple manner. We use a Magnet suspended in a pop bottle and acting as a Pendulum. In bringing other magnetic items (and yes, also magnetic-sensitive metal objects such as iron) near our simplified Magnetometer will cause it to swing. If allowed to swing not in the presence of an exterior force, it will only be affected by gravity and therefore behave as an ordinary Pendulum. With a disturbing force nearby (such as a test magnet) the Period of the Pendulum will vary and we have enough sensitivity with our instrument to notice, measure and calculate this.

Purpose : To construct a basic magnetometer and use it to determine magnetic field strength for various multiples of magnet sources at a given distance through variations in periodic motion of the pendulum in the device.

Purpose : To construct a basic magnetometer and use it to find the variation of magnetic field strength at various distances through periodic motion of the pendulum in the device.

172

Materials :

- Set of 2 strong small Neodymium button Magnets,
- Stopwatch,
- Clear Plastic Pop Bottle (2 L size best),
- Set of refrigerator and other small magnets to test (best to have a small set of regular button magnets),
- String,
- Scissors,
- Ruler,
- Tape,
- Glue or Clay,
- Slide Rule

Set Up & Construction of Device Plus Initial Calibration Measurements :

1. The reason for 2 neodymium magnets is that with opposite poles of the magnets on either side of the string being used it will hold itself in place without the need for tape or glue. If you have only one, then you have to decide how to attach it to the end of the string. – Note, however, one could use ordinary ceramic magnets (which, by the way, are the best test magnets).
2. Be sure to have cleaned the pop bottle. It is recommended to use a 2 L size for the magnetometer.
3. With the magnet(s) attached to the string, lower it into the pop bottle and allow it to hang about 3-4 cm above the bottom. This will determine the amount of string needed. Cut it a little longer than this since the excess will go into the pop bottle cap and be held in place by one of the following : tape, glue, or clay. Tape or Clay is recommended, since you may have to take it out to adjust it.
4. Fully assemble the magnetometer so that the magnet-based pendulum bob is hanging in the pop bottle and the pendulum is attached to the now closed pop bottle.
5. To test it, bring one of the test magnets near it and see if it reacts and begins to swing in a oscillating pattern as expected by a pendulum. Once in motion, move the test magnet away and let it swing naturally without any magnetic force influence.
6. (Observation : When there is a magnetic field near it and then not, do you notice any changes in the pendulum motion ?)
7. Remove the cap, string, magnet pendulum to conduct the calibration test to compare to for the natural motion of the pendulum.
8. Pendulum Calibration :
9. Use tape on the top of the pop bottle cap and attach it to the ruler.
10. Place the ruler between a heavy book and a table so that the pendulum hangs freely off the edge of a table.
11. Draw the pendulum back and let it go. At that exact moment start the stopwatch.
12. Note it is best to have a set number of complete swings (oscillations) of the pendulum and time these for whatever time it takes. Choose a consistent number, such as 8, 10, or 12, et al.

13. Do this at least 3 times and average the results for time. Note that time is not the same as Period. Period is the time for 1 oscillation.
14. Determine the Average Pendulum Period for the natural pendulum by taking the average time divided by the average number of chosen oscillations.
15. Check your results with the Period Formula for a conventional Pendulum to see how much it varies. If your variance is large, recheck your results and retest if needed.

Procedure :

1) Be sure to have done the prior Set Up and Calibration steps since this is where the Procedure begins.
2) With the Magnetometer reassembled (the cap, string, and magnet is put back on the bottle).
3) Be sure to have the bottle on a level surface and away from metal and magnetic objects so that it will only be influenced by the test magnets brought near it in the various Activities.
4) For the Activities each requires the measurement of the distance from the Pendulum to the outer bottle edge. It is best to do this with the Ruler centered on the pop bottle cap (which is atop the bottle) and looking down to where the edge of the bottle is. Record this value [C] in centimeters. It is a value added to all other measures of distance when called for in the Activities.
5) Note : It is also best to test brining the test magnets towards the assembled Magnetometer. The key is to notice a few things : First, does the Magnetometer work at all and what needs to be done to allow its proper operation?
6) Note : Second, how close does the test magnet need to be to allow for proper operation? This is critical since too much magnetic force will cause the pendulum to adhere to the side of the bottle when very near the test magnet. The test magnet must never be closer than a distance that allows it to disturb the pendulum's natural motion but does not entirely stop its motion.
7)
8) Activity 1 : Number of Magnets at a Fixed Distance
9)
10) In this Activity you will choose a given distance that will remain constant for the entirety of the Activity. To best determine this distance, take one of the test magnets and try various distances to see when they influence the pendulum. To test each distance, realize that you have to let the initial influence subside before further testing. In the end you want some influence. This may be anywhere from 2.0 to 6.0 cm from the bottle's edge. Note these distances do vary and depend on the relative strength of the magnets being used and are merely suggestions.
11) With the determined distance, be sure to record this value [D] which includes the constant distance of the pendulum from the bottle's edge in the Data Table.
12) There are at least 3 trials in this Activity (hence needing minimally 3 test magnets). It is best to have consistency in size and strength of magnets. Small Button versions are the best.
13)

14) In each trial, write down the number of magnets and then place that number of magnets at the chosen distance (recommended to have a marker there of some sort, since it will be used for all Trials). It is also best to start with the minimum number of magnets (1) and increase in each of the subsequent Trials.

15) For each Trial with the number of magnets in hand place them at the chosen distance position and then observe the motion of the Pendulum.

16) Time the Pendulum once you have an idea of its motion and count the number of complete oscillations it makes in the time measured.

17) Record the Number of oscillations [N] and the amount of time [t] for these oscillations.

18) Do this same Trial with the given number of magnets 3 times and record the results for Number of Oscillations and Amount of Time.

19) Average the Results and Determine the Period of the Pendulum for that given Number of Magnets for each Trial of Varying Numbers of Magnets. Note there are as many Periods as there are Trials for a given number of magnets – the trials of differing numbers of magnets are not averaged together!

20) In each case, there should be some variance from the natural motion of the pendulum and its period.

21) How do the number of magnets affect the Period? Could you reach a general conclusion (assuming each of the magnets being used has the same strength)?

22) Graph the results where the y-axis is the Period of the Pendulum and the x-axis is the Number of Magnets. Draw a best fit line and determine the slope of the line, if it appears linear. If not read on to the next step :

23) If the line is not linear, then graph the log value of the Period of the Pendulum by using the L scale of your Slide Rule vs. the log of the Number of Magnets and again draw a best fit line and determine slope. The slope in this case is the exponential relation of the number of magnets to the period of the pendulum.

24)

25) Activity 2 : Given Number of Magnets at Varying Distance

26)

27) In this Activity you will choose a given number of test magnets to use (2 or 3 may be best) that will remain constant for the entirety of the Activity.

28) In this Activity, you are to chose at least 3 distances. Use the initial distance used in the first Activity since it is probably the closest that the test magnets can come to achieve proper operation of the magnetometer. Then choose logical increments to this distance. They can be whole number or other types of multiples of the original distance (such as : 1.2, 1.5, 2, et al). Do not have distances that do not have any effect on the magnetometer (unless you make this is one of your personal goals for the exercise).

29) Recall that with each distance, be sure to record this value [D] which includes also the constant distance of the pendulum from the bottle's edge in the Data Table.

30) You do have the choice of recording the values as actual readings, such as x cm, y cm, etc or as multiples of the initial distance, hence the first is 1, the next may be 1.2, then 1.5, 2, and so on.

31) There are at least 3 trials in this Activity (hence needing minimally 3 distances). Note : It is best to perform each trial in a sequential incremental manner.

32)

33) In each trial, write down the number of magnets decided upon and then place them at each of the distances in turn for each of the Trials.
34) For each Trial with the test magnets at the trial distance now observe the motion of the Pendulum.
35) Time the Pendulum once you have an idea of its motion and count the number of complete oscillations it makes in the time measured.
36) Record the Number of oscillations [N] and the amount of time [t] for these oscillations.
37) Do this same Trial with the test magnets at the trial distance 3 times and record the results for Number of Oscillations and Amount of Time.
38) Average the Results and Determine the Period of the Pendulum for that given Distance for each Trial of the test magnet set. Note there are as many Periods as there are Trials for a given distance trial – the trials of differing distances are not averaged together!
39) In each case, there should be some variance from the natural motion of the pendulum and its period.
40) Graph the results where the y-axis is the Period of the Pendulum and the x-axis is the Distance. Draw a best fit line and determine the slope of the line, if it appears linear. If not read on to the next step :
41) If the line is not linear, then graph the log value of the Period of the Pendulum by using the L scale of your Slide Rule vs. the log of the Distance and again draw a best fit line and determine slope. The slope in this case is the exponential relation of the distance to the period of the pendulum.
42) Examining the Data Tables, how are the Periods varying – staying the same, increasing or decreasing with increasing distance? Why do you think so?
43) In the matter of both Activities, you can do other calculations – such as Frequency, which is the inverse of Period and can readily be found on the C1 scale of a slide rule when read properly.

Data :

Pendulum Calibration :

Trial	Number of Oscillations [N] (no unit)	Pendulum time [t] (s)	Pendulum Period [T] (s)
1			
2			
3			
Average			

Distance from Pendulum to Bottle Edge [C] : _____ cm

Activity 1 : Measure of Magnetic Field Strength of Increasing Number of Magnets at a Given Distance

Note : For any given distance, do the trial 3 times and average the results
  Be sure to recreate this Data Table for each Trial

Distance [D] : _____ cm
  Be sure to include the distance from the pendulum in the bottle to the outer surface!

Number of Magnets for Trial : _____

Trial	Number of Oscillations [N] (no unit)	Pendulum time [t] (s)	Pendulum Period [T] (s)
1			
2			
3			
Average			

Activity 2 : Measure of a Magnetic Field Strength for a Given Magnet at Varying Distance

Note : For any given distance, do the trial 3 times and average the results
  Be sure to recreate this Data Table for each Trial

Number of Magnets used for all Trials : _____

Given Distance for Trial : _____ cm
  Be sure to include the distance from the pendulum in the bottle to the outer surface!

Trial	Number of Oscillations [N] (no unit)	Pendulum time [t] (s)	Pendulum Period [T] (s)
1			
2			
3			
Average			

Calculations :

Be sure to use your Slide Rule!

Slope :

$$m = \frac{\Delta Y}{\Delta X} = \frac{\text{Dependent Variable}}{\text{Independent Variable}}$$

Average :

$$X_{ave} = \frac{\sum x_n}{n}$$

Period :

$$T = \frac{t}{n}$$

(T = Period for one oscillation in seconds, t = total time for n number of oscillations)

Frequency :

$$f = \frac{1}{T}$$

Pendulum Period Formula :

$$T = 2 * \pi * (l/g)^{1/2}$$

'l' is pendulum length, 'g' is acceleration due to gravity
g is taken as 9.8 m/s^2
'T' is the period of the pendulum measured in seconds (s)

Conclusion :

How did your Magnetometer function? What sort of relation did you find with a given magnetic field intensity with distance from a constant magnetic source? What sort of relation did you find when increasing the number of magnetic field sources at a given distance?

Activity #17
Wind Speed Electrical Energy Study Activity
Grade Level : High School
Math Level : Challenging

A Wind Turbine is any device that converts the kinetic energy of the wind into what is classically called mechanical energy (which is another type of kinetic energy). If the device is used to produce electricity, it is also called a Wind Generator. They look like a large fan typically with 3 or 4 blades.

Note that these devices have a long history and have been used long before the times of electrical energy production. Many of us know of the windmills of Holland, for example. These along with others in other parts of the world in different times have been used to grind grain and pump water. They date to Persia as early as 200 B.C. The first known example of one powering a machine comes from the windmill of Heron of Alexandria. Their growth started in the 7th century in the area of Sistan, a region between Afghanistan and Iran today. By the Middle Ages they show up in Europe.

Today there is an interest to harvest the wind to make use of it on large scales for electrical power production. There are not only the large scale ones, however. In sailing there are small models used for recharging batteries for auxiliary power.

The first to use the wind turbine as a generator to generate electricity was made operational in 1887 by James Blyth, a Scottish academic to charge batteries. He had done so in order to light his holiday home in Marykirk, Scotland. Shortly thereafter, a well known name in terms of motors and their parts, Charles F Brush develops the first automatically operated turbine for electrical energy production in Cleveland, Ohio.

The early 1900s saw a massive growth in the wind turbine industry. In Denmark in 1900 there were approximately 2500 windmills used for mechanical loads and used as pumps and mills and harnessing about 30MW of power. The largest of these at that time were 24m (79 ft) towers with 4 blade systems and the blades having an effective diameter of 23m (75 ft). In 1908, America had 72 wind-driven electric generators ranging from 5kW to 25 kW in power generation.

When it comes to the design of the wind turbine, there are two basic models : Horizontal and Vertical. This refers to the axis of rotation. The Vertical Type has its axis pointing away from the ground, basically straight up. It spins a blade-like system of sails in a round-and-around fashion like a merry-go-round. Here the generator is at its base on the ground.

The classic type we think of commonly is the Horizontal model that is like a pinwheel. The system must point into the wind to make the blades move and the generator is often right behind it atop the tower with the blades.

To measure the wind energy a quantitative measure called the Wind Power Density (WPD) is used and is a calculation of the mean annual power available per square meter of swept area of a turbine and is tabulated for various heights above the ground. The calculation involves the effect of wind velocity as well as air density. There are designations from Class 1 (200 W/m^2 or less at 50 m altitude) to Class 7 (800 – 2000 W/m^2).

The common one in use today are typically 3-bladed and pointed into the wind. They are computer controlled units. They are very high efficiency systems and can deal with tip speeds (the speed of the rotating blade) of 320 km/hr (200 mph) and have a low torque ripple. The blades are a light grey in color (to blend with clouds) and range in size from 20 m to 40 m in length (66 ft to 130 ft). The average tower is between 60 m to 90 m tall (200 ft to 300 ft). The blades typically rotate at 10-22 rotations per minute. At the upper end of 22 rpm the tip (end of the blade) speed is 300 ft/s (91 m/s). A common 80 m tall model with an electric generator can produce 1.5 MW. The generator is not merely a single gear system, but has a number of gears so as to maximize the output from the inputs given.

Outside of the large-scale ones, there are personal smaller ones that are sold in kits and can be installed as needed (permits and laws permitting such in a given area).

Though wind speed is important, so too is the size of the rotor blades. The power available to the blades is proportional to the square of the diameter of the rotor. In theory, if the turbine blades are twice as long, the power producing capability can go up by a factor of four (test this idea yourself in the Activity).

Today, there is an increased interest in the prospects of wind farms and generating electricity for remote and/or rural communities through wind turbines as well as being used for larger-scale usage by major metropolitan areas so as to relieve the load on the system and diminish the over-reliance on fossil fuel sources, such as coal.

In the Activity, you create a very small version of a wind turbine and connect it to a small model electric motor. Recall from the Electric Motor Activity (#13) that the electric motor is really a generator in reverse essentially. This is how we are using it here. To determine the power of the wind turbine created, we use a multimeter and measure the voltage and amperage it produces in a given set of conditions (wind speed, wind blade area, or number of blades – your choice).

Activity A :
Purpose : To determine the voltage output of a wind turbine (personally-made small scale model) as it varies with wind speed.

Activity B :
Purpose : To determine the voltage output of a wind turbine (personally-made small scale model) as it varies with the surface area of the blades.

Activity C :
Purpose : To determine the voltage output of a wind turbine (personally-made small scale model) as it varies with the number of blades.

Materials :

- Thin Cardboard (for wind turbine blades),
- Compass or can use lids to make circles to act as formats for the wind turbine blades,
- Scissors,
- Wire Cutter,
- Ruler,
- Rubber Band,
- Small model Electric Motor (small, about thumbnail size),
- Cork,
- Masking Tape (or Duct Tape),
- Clear Tape,
- Hand-held Hair Dryer with at least 2 speeds and best if a setting with no heat as well,
- Meter Stick,
- Multimeter,
- Alligator Clip ended wires (as long as possible),
- Stack of Books,
- Anemometer (useful but not really needed),
- Slide Rule

Safety Notes : When dealing with a personally made item Always exercise Safety. First and foremost, have parental permission and supervision in any part of the activity (set up and implantation of the procedures). Second, Always wear goggles for all parts. Third, Do not connect the wind turbine to any electrical source – the wind does the work to make the turbine go and it, in turn, generates an electrical current measured by your multimeter. Fourth, if at any time the system seems unstable – you must then stop the air flow immediately and let it slow down on its own. Do not stand near or in the potential path of any parts. When in operation, stand a safe distance from the system, even when taking measurements (the multimeter is connected by extended wires to the motor of the wind turbine).

Note : As with any project, it is best to have back-up supplies, such as additional rubber bands, enough tape, a second ruler, and so on. This is especially true in this Activity since there is a lot of work to be done on the blades for the wind turbine.

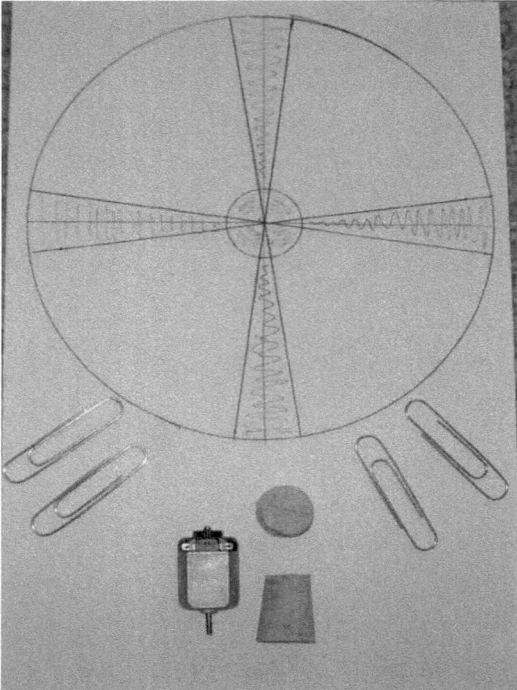

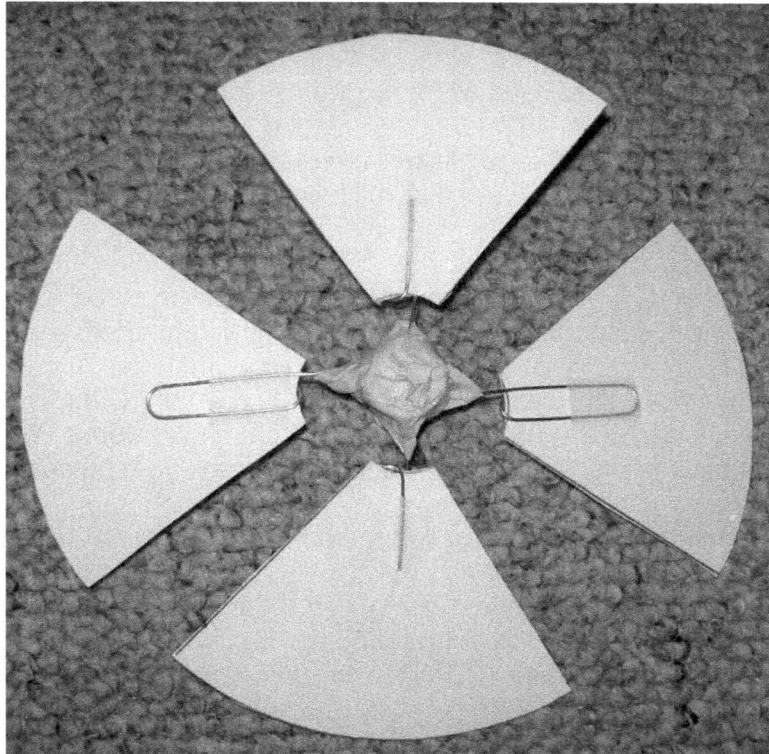

Set Up :

1. In all construction of the system and the operation of the Activity in the Procedure below as well – be sure to wear goggles.
2. This set up applies only once since the same system is used for all of the experiments.
3. The stack of books will become the tower for the Wind Turbine.
4. A ruler will project from it situated about 2-3 books from the top.
5. At the end of the ruler the electric motor will be held in place by a rubber band.
6. Do not connect the motor yet.
7. The first major construction is the blades. Decide if you are using a compass or lids to create a circular pattern to be cut out.
8. Once the outer outline is drawn, decide on the shape of the blades and use a ruler to draw the shape you have in mind.
9. Note : You have to be able to calculate the surface area of these blades. The formulae listed below are for common geometric forms, but may not be enough in this case.
10. Another alternative method is to place the form for the total surface area for the blades before cutting out on a piece of graph paper and trace it. Knowing the area of one of the squares, count the number of squares that the form covers (be sure to subtract the shaded areas that will not become part of the blades).
11. Record the information needed to determine the area of the blades.
12. You can prepare for the later Activity of the number of blades or area of the blades by completing all of this work here. If not, you will have to come back to this part and redo the steps needed to make the blades.

13. Cut out the blades and set them aside for the time being.
14. Take the paper clips (large seem to work best) and extend the outer loop of the clip so that it is a straight line. You need as many of these as you have blades (1 clip per blade).
15. Use the wire cutter and cut off about half of the extension (use your cork as a v sual guide to see how long it should be since at full length it will probably be too long for the cork).
16. Carefully put the paper clips into the cork.
17. Cut strips of masking tape and hold each of the paper clips in place.
18. Put each of the blades separately into the paper clip and tape them (with clear tape) in place.
19. Put a piece or two of masking tape across the back of the cork (that is attached to the stem post of the electric motor). It acts as an additional hold. Note : With various trials and adjustments between trials, you may need to put masking tape again on the assembly.
20. With the rubber band attach the motor to the end of the ruler with the stem of the motor pointed away from the ruler.
21. Now put the cork wind turbine blade assembly onto the step of the motor.
22. Make adjustments as needed (such as rotating the system to see if it spins free and clear, twist the blades so that they all point in the same direction (i.e. at the same angle)).
23. Use alligator clips and attach it to the multimeter and the other ends to the clips on the electric motor.
24. First have the multimeter set to voltage (and then later amperage). Be sure to know how to configure the system for this.
25. Note : The device will be measuring mV and mA.
26. Hand spin the blade system to see if it measures voltage (amperage). Recognize that a negative sign merely means that the voltage (current) is moving in the opposite direction from your set up (you can reverse it if you want).
27. Complete the Activity area set up so that the hand-held hair dryer is ready to go and test it first, of course.
28. One of the things to check is the operation of the system when the hair dryer is on low setting and at high setting. You can also determine where the distances should be for measurement taking. (Depending on the speed of the wind as produced by the hair dryer, it can be as little as 25 cm and as far as 125 cm).
29. Be sure to make adjustments as needed to the system as it operates. If it wobbles (the turbine system) be sure to stop and make adjustments. It is best to design a set up system made of materials that have a lower chance of causing problems (hence thin cardboard blades) and to maximize one's distance from the rotating system and to carefully monitor it when in operation.

Procedure :

1. Be sure to wear goggles.
2. Be sure to have parental permission and supervision.
3. For any and all Activities do not stand near the system while in operation. It is best to be in front of it where the wind is coming from (i.e. the hair dryer).
4. If at any time the system seems unstable, turn the hair dryer away from the wind turbine and let the turbine come to a stop itself.
5. As with any and all Electrical Activities the device you have constructed is not connected to batteries nor to any outlet. The only electrical device used is a common hand-held hairdryer. In fact, for those more adventurous, let the wind do the work, though one has to be patient and have the opportunity to use it.
6. **Activity A : Wind Speed vs. Turbine Power**
7. In Activity A, the Wind Speed is merely affected by a choice of two things.
8. First, you can take the hair dryer and change the setting when at a given distance (low to high, for example).
9. Second, another way to affect wind speed is to change the distance from the turbine (twice as far as it is initially, et al).
10. In any case, be sure to write down a description so that the relative wind speed can be understood (such as having a Low setting for trial 1 and 2, only the distance is 30 cm for trial 1 and 60 cm for trial 2). Note : Here the values for speed will be the inverse of distance ($\frac{1}{d}$).
11. Another way to determine the actual wind speed is to use an anemometer in each of the trials to check what the wind speed is (if one is available).
12. Note in the case of Activity A, the variable of wind speed will either be measured or relative (based on the inverse-distance of the hair dryer from the wind turbine).
13. For each of the trials measure the Voltage and the Current. These are used to determine the Power of the Wind Turbine.
14. Be sure to practice and know how to set up the multimeter and operate it properly to measure voltage or current. Note these will be DC.
15. For each trial do it 3 times and take the average value of the readings.
16. Calculate the Wind Turbine Power.
17. Graph the data of Power (y-axis) vs. the Speed of the Wind (x-axis). If the wind speed is relative use inverse-distance in place of anemometer readings.
18. Draw a best fit line and determine slope. (Note : In the relative speed case it will be with respect to inverse-distance).
19.

20. **Activity B : Blade Surface Area vs. Turbine Power**
21.
22. With the system set up and operational in this Activity you need to choose as constant wind speed and distance to operate from. Be sure to record these numbers.
23. As noted in the Set Up it is a good idea to have a series of turbine blade sets ready to go.
24. Measure the surface area in the manner noted in the Set Up and record it for each of the Wind Blade Sets in the Data Table.
25.

26. It is probably best to choose a direction, such as smallest to largest blades when conducting the trials.
27. For a given trial for a given wind blade size perform the experiment 3 times, measuring the Voltage and the Amperage for a given time. Average these results.
28. Now change the Blades with ones that are larger or smaller for a next trial.
29. For each of the trials measure the Voltage and the Current. These are used to determine the Power of the Wind Turbine.
30. Calculate the Wind Turbine Power.
31. Graph the data of Power (y-axis) vs. the Wind Turbine Blade Area (x-axis).
32. Draw a best fit line and determine slope.
33. Is there a relationship?
34.

35. <u>Activity C : Number of Blades in Turbine vs. Turbine Power</u>
36.
37. With the system set up and operational in this Activity you need to choose as constant wind speed and distance to operate from. Be sure to record these numbers.
38. As noted in the Set Up it is a good idea to have a number of blades in various set or series of turbine blade sets that are ready to go.
39. In this Activity, it is merely the number of blades, so in going from one trial to the next, you may have to take out the paperclips with the blades and then rearrange them so they are again balanced.
40. Note that if you want to maintain consistent surface area of blades, the sets with a smaller number of blades need to be larger blades, while the ones with the most blades will have the smallest blades.
41. This means that for each set you should measure the surface area in the manner noted in the Set Up.
42. It is probably best to choose a direction, such as the least number of blades to the largest number of blades when conducting the trials.
43. For a given trial for a given wind blade size perform the experiment 3 times, measuring the Voltage and the Amperage for a given time. Average these results.
44. Now change the Blades with ones that are greater in number for a next trial.
45. For each of the trials measure the Voltage and the Current. These are used to determine the Power of the Wind Turbine.
46. Calculate the Wind Turbine Power.
47. Graph the data of Power (y-axis) vs. the Number of Blades on the Wind Turbine (x-axis).
48. Draw a best fit line and determine slope.
49. From the slope, is there a relationship?

Data :

Activity A :
Wind Turbine Power with respect to Wind Speed

Trial : Wind Speed or Distance	Voltage reading for Wind Turbine	Amperage reading for Wind Turbine

Note : For a given trial conduct it 3 times and take the average for voltage and current measurements.

Activity B :
Wind Turbine Power with respect to Blade Surface Area

Trial	Wind Turbine Blade Surface Area
1	
2	
3	

Note : For a given trial conduct it 3 times and take the average for voltage and current measurements.

Trial : Blade Surface Area	Voltage reading for Wind Turbine	Amperage reading for Wind Turbine

Activity C :
Wind Turbine Power with respect to the Number of Blades

Trial : Number of Blades	Voltage reading for Wind Turbine	Amperage reading for Wind Turbine

Note : For a given trial conduct it 3 times and take the average for voltage and current measurements.

Calculations :

Be sure to use your Slide Rule!

Power Calculation :

$$P = V * I$$

Slope formula :

$$m = \frac{\Delta Y}{\Delta X}$$

Conclusion :

Of the various Activities noted here, which had the greatest impact in the Power generated by the Wind Turbine? In a given Activity, what was the general outcome of the Power generated by the Wind Turbine? Can you think of other Activities (such as having a carefully measured angle of the wind turbine blades?)

Determine the Specific Heat Capacity of Water
Grade Level : High School
Math Level : Calculating

Specific Heat Capacity of Water Activity

Consider for a moment the old adage : 'A watched pot never boils'. Where does it come from and why do we think that? The nature of water is the reason. It heats up very slowly. In fact, if we had two spoons in that same water as it heated, say a metal one and a wooden one, we know which to handle with care. The metal one! Why? In all of the cases, the slowly heating water, the almost as slowly heating wooden spoon and the more rapidly heating metal spoon is due to what is known as the specific heat capacity of the material.

Some of the common factors we might consider beyond composition are the amount of matter, the initial temperature of the material along with the temperature of the surroundings , along with the color of the material (why we often wear dark colors in the winter as opposed to the summer with lighter colors). Even the state of matter will determine the rate of heating it. Take for example H_2O. It is ice in a solid state, water in a liquid state and steam in a gaseous state. In each of these states each has not only a different amount of energy per unit amount, but also a different specific heat capacity based on the arrangement of the molecules themselves. In fact, there is another consideration in going from one state to another and that is a whole other lab consideration not examined here. All of the aforementioned are not specific heat capacity itself. This property is directly connected to the nature of the material itself.

The way to determine the physical property of the amount of energy per unit mass and change of temperature of a substance is known as the specific heat capacity of a substance.

$$Q = mc\Delta T$$

The enegy, Q, is measured in either Joules or calories (depending on the chosen units). In fact, a calorie is defined by this latter formula as coming from water. The specific heat capacity of water is :

$$c_w = \frac{Q}{m_w * \Delta T}$$

For water, for each gram of substance, the change of $1°C$ requires what is defined as 1 calorie (4.186 Joules). This value determination is the goal of the Activity.

The energy, Q, does not only go into the change of temperature, which is directly related to the translational motion of the particles of the substance. It can also go into the rotational motion of the particles, as well as the internal vibrations of the particles bonds as well.

To determine the specific heat capacity of a material the calculation is rather straightforward. Merely measure the amount of energy needed to change a given amount of mass by a specified change in temperature. The ratio (noted above) of the energy to the product of the mass and change in temperature is the specific heat capacity of the material. Metals tend to have values less than one. The lower the value, the easier it is to change a given mass by a given amount of temperature change. Say, for example, a known metal has a specific heat capacity of $0.5\frac{Cal}{g*°C}$. this means that it requires 0.5 calories for each gram of substance and its change of temperature of 1°C.

Compare this to water, which is $1\frac{Cal}{g*°C}$ = 4.186 $\frac{J}{g*°C}$, indicating it needs twice the amount of energy for this same outcome. This might be where the old adage noted , 'a watched pot never boils' may arise from.

The energy value, Q, in the equation tells how much is needed in one consideration or can be given off by a substance at a given temperature too. This is important to this activity since to find the specific heat capacity we employ the unknown metal in question and water that we know plus the critical basic idea of conservation of energy :

Heat Lost = Heat Gained

Sum of all Heat Q = 0
$Q_W + Q_M = 0$

The heat lost by a material, here we are using an electric kettle/boiler (best to use a travel heater that can run on AC) and we assume that all of its energy is given to the water. We use this idea to find how much the temperature of some given amount of water at a measured initial temperature changes during a measured time interval. Each of these becomes a point on our graph (energy used versus mass and temperature change). The slope of the best fit line in this case reveals the specific heat capacity of water.

Why is the study of heat so vital? The obvious answer is nature and weather itself! Land and Water heat at different rates and the angle of the Earth's tilt with respect to the Sun plus its rotation creates a very complex system that results in different rates of heating along with the fact that the incoming radiation is partially reflected into space before even entering the atmosphere (about 30%). Different materials absorb at different rates and the energy can be redistributed, reabsorbed and retransmitted in the infrared (concerns about global warming due to materials even in the air such as CO_2, et al). The amount of water vapor, dust, along with the angle the sun strikes the surface, and the texture, and composition of the terrain plus the wind speed will affect the surface temperature. All of these affect the atmosphere's energy content, hence affecting pressure and pressure differences is the driver of the winds along with other factors too. This drives the weather!

Closer to home, heat and its transfer is critical in immobile structures (homes, buildings, et al) and even in mobile structures (cars, trucks, engines, et al). In buildings it is the retention of heat, particularly at cold times that is of importance. Yet when it is warm outside, the retention of cold air and the cooling of the air is critical. The measurement, dissipation, and control of the heat is very important. Even with a computer and its cooling fans and other appliances, such as refrigerators, and those that use heat like toasters, ovens, and coffee pots, heat once again becomes a factor of importance.

This is where Specific Heat Capacity comes into consideration. Each material has its own value for the amount of energy needed to change a given mass unit by a given amount of temperature change. The higher the value (such as with water at $4.186 \frac{J}{g*°C}$ means it takes a great deal of energy to change the temperature of water. Compare this to most metals which it requires less energy to change its temperature. For example aluminum has a value of $0.897 \frac{J}{g*°C}$ while common wood has a value of $1.70 \frac{J}{g*°C}$. both are less than water hence heat up faster, but notice how much less the metal is as compared to the wood. All the more reason why a metal spoon in the hot pot of water is dangerous to touch and very hot very quickly as compared to the wooden spoon in the same pot (do not try this).

This physical property of matter, the specific heat capacity, is but one of many ways to examine matter. Other ways can include density, rate of expansion when heated, electrical conductivity, and the like. These properties, such as the specific heat capacity, can be used to determine its value or usefulness in various applications, such as when employed as a conductor or an insulator.

The study of heat is a branch of physics called thermodynamics . What of the definition of heat and the units of heat? In physics, heat is considered the energy transferred from one body to another due to thermal contact due to differences in temperature. Naturally heat transfers from a body of warmer temperature to one of cooler temperature. When bodies are at the same temperature they are said to be in thermal equilibrium. When they are not the process of heat (energy) transfer can take place.

The examination of heat looks not only at appliances and buildings, but also weather, the interior of the Earth, and other bodies in the universe such as the Sun and other stars. Heat is a measure of the quantity of energy of a system, hence there is no 'cold' per se, merely the absence of heat as compared to another body. Also the Heat or Energy can only transfer via one of 3 known means : Convection, Conduction, and Radiation. The units of heat are : Joule, calorie, kilocalorie. In fact, the factor noted before is the conversion for calories and joules : 1 calorie is 4.186 joules. Note that these are not the food calories we consume, each of these is 1000 of the base unit calories, hence the food ones are often written with an upper case 'C' as Calories. We could then convert our food calories to kilojoules by multiplying by the conversion factor. In many European countries, cans and boxes of food do not note the Calories, but instead the kilojoules instead.

Heat is something which may be transferred from one body to another according to the 2nd law of thermodynamics (i.e. the noted natural process and via one of the aforementioned means or if it does go from a place of lower energy to one of higher energy, it requires the input of work).

Returning to the goal of the activity : Heat will be transferred to the water and we want to determine its ability to take in and manifest that energy due to its mass and the change of temperature it undergoes. This demonstration of energy is thermal and is illustrative of its nature. From it we can determine the property of specific heat capacity. One of many heat factors we need to know to examine any of the other dimensions of heat from the atomic to the macroscopic in all of the noted fields.

Purpose : To determine the specific heat capacity of water

Materials :

- Tap Water,
- Thermometer (lab quality),
- Stopwatch,
- Home model (typically travel) electric kettle or boiler,
- Measuring Cup with metric readings,
- Graph Paper,
- Ruler,
- Slide Rule

Notes :
1) In handling hot water be sure to have permission from adults or are supervised by adults if necessary plus exercise caution.
2) In this Activity we are assuming that tap water is free of dissolved materials and has a density of 1 g/cc. For better results, use distilled water.

Procedure :

1) Record the Power Rating for the electric kettle/boiler in the data table (P) which is found on a label on the device or on the box it came in.
2) Fill to its normal level the electric kettle/boiler and have it run through its time to boiling the water and pour it out (this is to heat up the system).
3) In the time of the kettle/boiler heating, set aside a measured amount of tap water in a measuring cup.
4) Record the amount (V) [measured in mL which are cc] and record this as the amount of grams in mass (M) too since M=D*V, where D=1g/cc here.
5) Also for the water, measure the initial temp (T_i) [in °C] and record it on the data table. Note for all initial temps, record it once in the device, but it is not plugged in or activated. Let it sit for about 1-2 min in it then record the initial temperature.
6) Note that you must operate the system as noted in the directions and do so safely. Do not put things in there while it is active. After the temperature reading, the thermometer is removed and the system is then plugged in and activated. Do not do this yet, read through all of the directions first. Keep in mind you need to time the time for heating the water to boil. It may be a good idea to wear goggles (have adult permission and supervision).
7) Depending on the size capacity of the kettle boiler you will use 4 different amounts, starting with the smallest amount, for example 400 mL and going by 100mL per trial up to 700 mL in the final trial. The amount depends on the size of the electric heater you are using. Choose the best range for this.
8) In each case record the mass and the initial temp of the water used.

9) Each of these amounts is a Trial on the Table below. Once a trial is done, pour out the water and let the electric heater sit for a few minutes. Then repeat the process by putting the next amount in and recording the values as noted in the following directions. Be careful with hot water.
10) For each trial let the kettle/boiler operate where it heats up to boiling. It will be assumed that the final temperature is 100°C.
11) For each trial, measure the amount of time (t) [measured in sec] with a stopwatch it takes for that given amount of water to heat to boiling (predict, how should time change as mass is increased for each trial?) and record it in the data table.
12) Compute the amount of Energy (Q) needed to heat the water to boiling due to the electric kettle/boiler and the amount of time needed. This calculation involves the system's power (P) and the amount of time (t) needed to heat the water.
13) Convert the Energy (Q) from Joules to calories!
14) Compute the quantity mass*change in temperature (m*ΔT) and call this the mass factor (N).
15) Graph as coordinate points the mass factor and the Energy needed for that factor (N, Q) this is (x,y).
16) Draw a best fit line through these 4 points and compute the slope of the line (c). The slope is your determination of the specific heat capacity of water!
17) Determine the percent error from your experiment with the accepted value.
18) Though the Slide Rule is a recommended tool, all of the calculations can be done with a graphing scientific calculator or the use of a spreadsheet program. In these calculations you have to generate a table of data, graph it, and then find the slope and/or equation of the best fit line for the data. Other formula calculations can be done with these tools as well.

Data :

Power Rating of kettle/boiler : (P) = _____ W

Final Temp. (T$_f$) = 100°C

Trial	V (cc)	m (g)	T$_i$ (°C)	ΔT (°C)	t (sec)
1					
2					
3					
4					

Trial	Mass factor N (g*°C)	Energy Q (cal)
1		
2		
3		
4		

Calculations :

Be sure to use a Slide Rule!

General Equation of Specific Heat Capacity for a material :

$$Q = m*c*\Delta T$$

Change of Temperature : $\Delta T = T_f - T_i$

Mass Factor : $N = m*\Delta T$

Energy : $Q = P*t$

Solving for Specific Heat Capacity (c) :

$$c = \frac{Q}{m*\Delta T}$$

$$c = \frac{\Delta Q}{\Delta N}$$

Slope : $m = \frac{\Delta Y}{\Delta X}$

water has a specific heat capacity of : $1\,\frac{Cal}{g*°C} = 4.186\,\frac{J}{g*°C}$

- use the above value for converting from Joules to calories

Percent Error : (use the value noted here)

$$\%E = \frac{[Experimental\ Value - Accepted\ Value]}{Accepted\ Value}*100\%$$

Conclusion :

The matter is clear here, what was the value obtained and how does it compare to the known value? If different, can we find and control some of the issues at hand and can there be refined measurements?

Notes :
As can be seen there are a number of assumptions being made here, but it should provide a reasonable and safe outcome to examine this idea.

Activity #19
The Efficiency of a Complex Electrical & Mechanical System
Grade Level : High School
Math Level : Challenging

Efficiency is simply the Ratio of Useful Work Output to Work Input. It is a natural consequence stemming from the Laws of Thermodynamics. Often in our Activities we have assumed a perfect transference of energy, such as Potential Energy to Kinetic Energy, but there is some loss of energy to heat and sound (such as the sound of a rolling marble and the small increase of temperature of the surface it rolls across).

This Activity is much like a Rube Goldberg device, though far less sophisticated, where there are a series of steps. We use a light source (perhaps a desk lamp or a flashlight – torch to United Kingdom users, or the Sun itself) to power some solar cells which in turn is connected to a small electric motor which is attached to a pulley with a string attached to a bag of masses (marbles perhaps) to lift them a given distance.

Each step is its own system we have looked at in some manner in other Activities, such as the light striking the solar cells or the lifted marbles being then propelled down a ramp, but here we have combined these.

In each case there will be Work Input by some energy source and Work Output. The ratio of Output to Input ideally would be 100%, but in reality, will be less as our results will show. Better still, in combination the final efficiency of our system will be the combined product of each of the steps multiplied together. Test this idea for yourself.

Efficiency is a very critical component of our everyday lives both in real measured amounts and in the processes we engage in – such as when we go out in the car we do several errands and connect the steps in a loop so as not to do multiple trips or re-crossing our path. When it comes to mechanical and electrical processes we find ways to increase efficiency – such as lubricated joints, use of ball bearings in motors, and the like to reduce friction, hence reduce heat therefore wear and tear on systems and convey the energy from one system to another more effectively with less losses. The better we do this, such as turning the stored energy of coal into steam then into electricity that powers our homes and factories, then the less waste there is.

Purpose : To measure the Work Input & Work Output of various subsystems of a complex system (electrical and/or mechanical) to determine Efficiency at each step of the system with its energy transformations.

Materials :
- Bag of Marbles,
- Ruler,
- Meter Stick,
- Batteries (if needed for the flashlight – if using a flashlight),
- Known Lumens value Flashlight (the higher the better, 1200+ best) OR,
- Light Bulb (75 W or 100 W) and Desk Lamp (best choice - see Note),
- Solar Cell(s) (the more used the better the system – 3 is good),
- Small Model Electric Motor (1.5 V – 3.0 V, can operate around 80 mA),
- Alligator Clip Wires,
- Pulley that can attach to motor (or one that can be made – see directions for Activity 4),
- Lab Quality Thermometer (Note : instead can use a multimeter with thermocouple),
- Mass Scale,
- Multimeter (possibly with various applications *),
- Tachometer,
- Oven Pad,
- Plastic Crate,
- Styrofoam Sheets (to cover in 2 layers the plastic crate), OR
- Cooler – (Styrofoam is good, small Plastic one with drain plug best),
- Light Socket with Wires and Plug,
- Timer,
- String,
- Tape and Duct Tape,
- Scissors,
- Goggles,
- Stack of Books,
- Rubber Bands,
- Slide Rule

Note : You must have parent permission and supervision in this series of Activities. Be sure to always act with caution and care. Throughout the process wear goggles, know how the equipment works, test materials, be patient, and be willing to redo the tests as needed.

Note : This set of Activities are a series of steps that begins with a light source (one of the following : the Sun, a lamp and light bulb system, or a very strong flashlight) that activates a set of solar cells which in turn are connected to an electric motor which turns a pulley with an attached string that lifts a small mass (small bag of marbles). In each step the energy going in (the work in as it is known) is measured / calculated and the energy out (called the useful work output) is determined. The ratio of the latter quantity to the former one is the efficiency of that step of the situation.

Note : This Activity has a number of steps and it may be best to test each individually to see it in operation and then join the subsystems to have a continuous stream of actions. Also this means you may have to run a number of trials to have it operate correctly.

Note : If using the lamp and light bulb system (hence the air calorimeter) be sure that your design does not overheat any surface. Exercise caution and let the system cool after operating for a given time. Do not place anything in contact with the bulb, especially when in operation.

Note : In the case of the desk lamp, it may be best to have a base socket for the bulb (see photo) and a plug that can attach (see photo). This like the other components of the activity require parent consent and supervision. This is best as compared to the desk lamp since it sits low to the ground, hence you may not need stacks of books and the like when using the air calorimeter.

Note : The air calorimeter has two choices : Either construct a make-shift system or use a sufficiently large cooler (the crate can be plastic or Styrofoam). The make-shift one uses a plastic crate (see photo) that has at least one layer of Styrofoam (can have two, one inside and one outside). In any case the key is this : when the light is positioned in the air calorimeter the active bulb must have clearance so as not to overheat any surface. Clearly do not touch the active bulb and let it cool sufficiently before moving. The goal is to have a system that can be 'closed' off and a temperature-measuring system is inserted and read for a short period of time as the air temperature changes (such as with a thermometer or using a thermocouple attached to a multimeter).

Note : * The recommended Multimeter has the conventional applications such as AC & DC measures for voltage, current, as well as for Resistance, but can also have a Photometer for light intensity (and is actually needed if using a strong flashlight as noted above). The model : Mastech MS 8229 is very well suited for this.

Note : If using the lamp and bulb system, you do not need a photometer in your multimeter. The light output will be determined indirectly with an air-calorimeter noted in the directions below.

Photos of Set Up and Other Components :
Note that the last photo of the Air Calorimeter is a work in progress

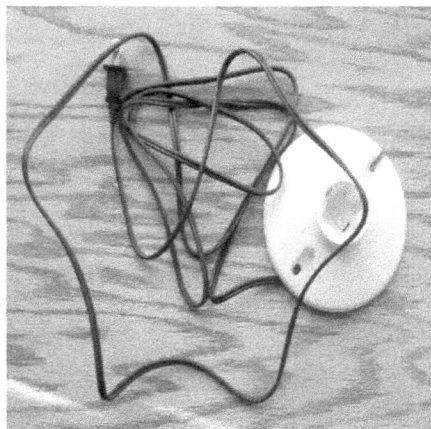

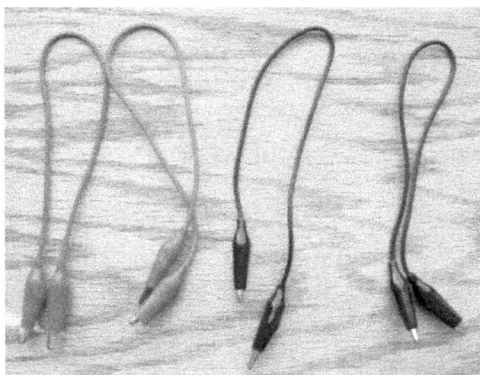

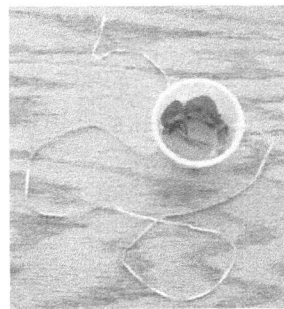

Procedure :

- **SET UP :**

- Note : With all of the Activities noted here – Always have parental permission and supervision. It is best to wear goggles. Always employ safe procedures.
- Note : All the Set Up and Measurement Procedures only set up the materials and take measurements, and in each section do the calculation for power in each section then in the Data Analysis section following all of these is where the efficiency for the data is addressed.
- This means that you may have to read the directions in a manner that satisfies what you are doing and where you are at. For example, if you had done Activity 1B (as an example illustration), then there is no need to read 1A nor 1C.
- Next you can then do computations for 1B in the Data Analysis OR continue in sequence for each of the Activities (2, 3, etc) until set up and measurements are completed.
- Note : You may have to do repeated trials and be sure to test each step of the path before having it operate. It is best to look at each component as a step. –
- In the case of multiple trials, it is often best to do a given trial as many as 5 times. Then eliminate both the highest and lowest values and then take the average of the other 3 where this average acts as the intended data measurement value.

- NOTE : Your light source determines which Activity 1 (A,B,C) you do :
- Choosing the subset of Activity 1 is determined by how you are able to measure the light source being used :

- If you have a Photometer, then you can do any of them,

- If you do not have a Photometer, then you must do Activity 1B,

- If you use a very strong flashlight, then start with 1A,

-
- If you use a light bulb and lamp, then start with 1B (recommended),
-
- **Set Up of the Air Calorimeter (used only with Activity 1B):**
-
- Note : The initial Air Calorimeter is shown in the set of photos and is the last one. Note that this is not complete, yet.
- What is needed is an open grate crate with thin Styrofoam sheets (can use foam-core poster board instead).
- The Styrofoam is cut so as to line both the inside and the outside of the crate along the plastic.
- It can be held in place by duct tape.
- Note to leave enough to make a 'lid' as well.
- Special Note : If this takes too much time, money, resources, one could instead simply use a Styrofoam Cooler in place of all the aforementioned items!
- An even better choice is a small plastic cooler with a drain plug – the need for a hole is because a thermometer will be inserted to measure the change in air temperature with a lit light bulb inside.
- Returning to the Air Calorimeter Set Up :
- With either the Styrofoam-lined crate or the Styrofoam Cooler, now invert it and pick a corner through which either a Thermometer will be inserted OR where the Thermocouple will be inserted. It is best to make a hole with a thin pencil or better still a knitting needle if available.
- Re-invert the box and use a piece of Styrofoam to construct a partial barrier that can be held in place with tape so that when the Thermometer or Thermocouple is in the Air Calorimeter it is not going to be in direct line of light from the light bulb which will sit opposite the temperature probe slot at the 'top' of the air calorimeter.
- It can be seen that there is a reference to the 'top' in a quotation-mark manner. This is because ultimately the box will be inverted and placed atop the 'lid' which lies flat on the ground and on top of the 'lid' in the center of it is the light bulb in a small light fixture (see photos) and connected with an appropriate electrical wire system so that it can be plugged in.
- The first main measurements to make here are the Length (L), Width (W), and Height (H) of the interior of the Air Calorimeter space. Record these measurements. Measure these to the nearest $1/10^{th}$ of a centimeter.
- This completes the Air Calorimeter Set Up and initial measurements –
- The Air Calorimeter needs the Volume (V) calculated from it and when coupled with the Density of Air (ρ_A) will yield the Mass of air in the Air Calorimeter (m_{air}) (do this calculation).
- Use this Air Calorimeter calculation with Activity 1B next
-
- If you use the Sun, then start with 1C,
-
- **The goal of Activity 1 is to measure the power of the light source, which will be the Input Work for Activity 2**
-

- ## **Activity 1A : A Strong Flashlight as the Light Source**
-
- Note : In order to use a flashlight, it must be a very luminous type, such as 1200 lumens or more. These can be expensive, so it is recommended to do Activity 1B or 1C.
- Note : If using a flashlight, it is best to only use an LED type and not other luminous source lamps which can be very hot. The flashlights described here are strictly battery operated and LED type and this is the only type recommended.
- Note : Since most, if not all very luminous lamps are LED types, it would be very difficult at best to use a simple air calorimeter, so the best measurement system would be to use a photometer to determine lux output of the light and then to translate this into Power.
- Note : The fastest way to test whether or not you can use this path is to quickly test your solar cells – connect them to the motor and see if this type of light can power the system. If it cannot, then you need to try either another method or find still a brighter light (if accessible – note that the more luminous types cost more)
- The first thing to measure is the Power Input into the light source.
- Use the Multimeter to measure the batteries both for Voltage (V_{1A}) and Current (I_{1A}) when not connected to the lamp. Be sure to know how to operate the multimeter to take these measurements. It is best to have the batteries connected in the same manner as they are in the flashlight (in series essentially) and measure them operating together. Record these results. The results from these measurements will be calculated and are considered Work Input #1.
- Now to measure the flashlight output and translate the light output into Work Output #1 (which in turn acts as Work Input #2).
- The most effective measurement of measuring the light is to use a photometer.
- Have the flashlight positioned so that it has an effective cone of light and the multimeter photometer sensor will be centered facing it at 1.0 m distance.
- The key to the measurement is to have as little outside light as possible, so you must find a means to read the meter (most digital have illuminated screens) such as drawn shades, etc.
- The reason for 1.0 m distance is the fact that the lux (lx) meter will be measuring lumens/meter-squared (lm/m^2)
- Record this lux measurement (L_{1A})
- Move on to Activity 2
-
- ## **Activity 1B : A regular Electric Lamp as the Light Source**
-
- With the Set Up completed first for the Air Calorimeter (see above) now we can measure indirectly the Power Input and Power Output of the Light Bulb as a Light source in our Activity
- Note as with any bulb, do not touch it when active nor handle it for some time after it is has been active since it is hot. Do not place objects on or too near the active light bulb. Be sure to have parental permission and supervision.
- First measurements are to determine the power going into the lamp.

- We can simply take the measure of the light bulb, say 75 W (a good recommendation), as the measure of the input power to the system.
- This value will be considered the Work Input #1 (W_I) and can be written on the table. If you are not measuring the light bulb's input power directly (requiring a specialized multimeter) now move on to the light output measure as work output #2.
- Light output as Work Output #2 Measurement :
- The output power can be measured either directly, using a Photometer as we did in the flashlight case (only here we use a light bulb), so if this is your method, read the directions in section 1A and replace flash light with the light bulb and do the same thing.
- The other measurement method is indirectly by measuring the heat the light generates and use this to subtract from the input power to find how much useful power is turned into light.
- Side Note : It may be a good experiment on its own to try both methods to see how similar or different these values are and consider why they are different!
- Indirectly measuring Power Output of the Light Bulb using the Air Calorimeter :
- Note that the Air Calorimeter must be completed (see directions above) for this next section.
- Here the bulb is in the socket and the Air Calorimeter is placed over it.
- Initially do not activate it.
- Use either a lab quality Thermometer inserted through the hole in the top (note that it needs to have a barrier to block the direct light upon the thermometer, such as a piece of white poster board or Styrofoam) or you can use a thermocouple attached to a multimeter for temperature readings.
- Take an initial temperature reading with the light off (T_i).
- Now start the timer, and take temperature readings every 2 minutes for at least 10-16 minutes total time (t).
- At this point you have enough measurements to be able to compute the amount of energy given off as heat by the light and then with the time (and proper conversion factors) convert this into watts.
- This is done by creating a graph of Temperature (y-axis) versus Time (x-axis), draw a best fit line and determine slope.
- With the slope you can now determine the heat energy of the system by using the specific heat capacity of air and the volume of air with this slope value – determine this power value.
- Note : This is NOT the power output of the light that we are going to use. We need to take this value and subtract from the Power Input value for the light and the difference is the power output of the light! This difference is the power value for the light we are going to use.
-
- **Activity 1C : The Sun as the Light Source**
-
- We have two choices here :
- Either make an assumption as to the power output of the Sun OR
- We can determine the value of the power output of the Sun by doing the Solar Constant Activity noted in the book.

- Note that the Efficiency in Activity 1 cannot be computed since the input is considered to be the same as the output (though not technically true, one could do the computations of how much of the Sun's energy is given off as light as compared to the amount of energy generated by fusion in the core, but that is another matter)
-
- Regardless of which of Activity 1 (A,B,C) you have done, now move on to the rest of the Activities in sequence :
-
- **Activity 2 : The Light Source acting on the Solar Cell**
-
- Regardless of which light source you have used, you now have a Power or Work Output #2 for a given light source to examine in Activity 2.
- This value will now act as Power or Work Input #2 for Activity 2.
- That logic will continue for each of the Activities, where the Work Output from the prior step becomes the Work Input for the next step.
- When it comes to Efficiency for a given step, it becomes the ratio of the Work Output for that step over the Work Input to that step.
-
- Now we are going to examine Power Output#2 for Activity 2, which comes from the use of one or more Solar Cells.
- Note : It is best to use 3 solar cells rated at 1.5 V and connected in series with each other for later power needs of the electric motor.
- With the Solar Cells connected properly and receiving the light from the light source, now use your multimeter to do the following :
- First, with proper settings on the DC Voltage, and connections of the cells to the meter (it is connected in parallel with the cells) measure the voltage of your solar cells with the light acting on them.
- Record this voltage value (V_{SC})
- Second, now reconnect properly the multimeter so as to measure DC Current (it will be in series with the cells) and be sure to have it on the proper setting so that the circuit does not blow – it is best to begin at the highest setting and then turn down to incrementally lower settings and noting the values along the way.
- Record the current value (I_{SC})
- You can readily calculate Power Output#2 here and do so, but can easily continue to the next Activity (3) where this acts as Power Input#3
-
- **Activity 3 : The Solar Cell acting on the Electric Motor**
-
- From Activity 2 we now have the Power Input#3 (which is Power Output#2) for this Activity.
- Now connect the Solar Cells to your Electric Motor.
- Test it to make sure it operates correctly

- Note, it is best to have a stand for the Electric Motor. A good idea is a stack of heavy books where a ruler is sandwiched somewhere near the top of the stack and jutting out a couple of inches. The Electric Motor is then held in place by a rubber band or two on the motor. Be sure that as little vibration is going on as possible when it is in operation. Also do not run the motor too long in any of the Activities so as to prevent overheating.
- Once the motor is operational connect small cut-out thin poster board circular disk (about ¾" diameter) to the axle of the motor so that it rotates.
- Measure and record the diameter of this disk in centimeters (d)
- Use the black reflective tape that comes with your Tachometer and attach a small piece to one edge of the disk –
- And to promote balance, attach a small piece of masking tape of approximately the same size and mass to the other edge of the disk so that it rotates in a balanced fashion
- Before using the disk with the tape on it, you need to measure its mass (m_d) and record this value in the table.
- Let the motor operate and use the Tachometer to measure the angular speed of the system. (ω)
- Note : We are going to assume that the motor has 'no load' on it, though there is one, hence this will affect the outcome slightly. In our next set of measures we are placing a significant load on the motor, though.
- Now Use the mass, diameter, and angular speed to determine the rotational kinetic energy of the system (KE_R)
- In the calculation section we are going to use this calculation for rotational kinetic energy as the power of the system to determine efficiency.
- Note : We are then going to make some critical assumptions :
- We will assume that the rotational kinetic energy started at zero
- We will assume that in each moment of time the rotational kinetic energy is constant
- With the value of rotational kinetic energy determined, we can then say that the system has this output of energy per unit time (1 second) and hence this same number of watts as the Power Output#3
-
- **Activity 4 : The Electric Motor acting on the Load**
-
- The Power Input#4 is the same as the Power Output#3 from above
- Now to the Electric Motor attach a pulley system (solid rotating disk – which may be purchased, but if unavailable it is best to use poster board cut into 2 disks approx. 1-2" diameter which are glued onto a central foam-core poster board disk of a smaller diameter (½" to 1.5" depending on the outer disks) leaving at least ½" to the outer rims of the make-shift pulley) – On the pulley you make as well you might want to cover the foam-core area with a thin piece of masking tape.
- Next have a piece of string less than 1.0 m in length but it should be longer than 50 cm.
- The length will be affected by how far in the air the motor is based upon the use of the stack of books as noted in the prior Activity.
- Test to be sure the pulley system is operational
- With an operational system we need to do a few measurements :
- Measure the diameter of the inner foam-core disk (ds) and record this.
- Measure the mass of the entire pulley disk system (mp) and record it.

- Now attach the string to the pulley and place a small mass on it (ideally use a plastic sandwich bag with a marble or two)
- Again test the system
- With further tests continue to add masses (marbles) until the system can still lift it but it requires some effort.
- Note : You may want to have back-up pulleys and be sure to keep the tension on your pulley system (a piece of masking tape may help) if needed
- When the necessary mass is in place, measure the entire mass (M) on the scale and record this value for the load of the system.
- Before activating the motor, be sure to measure the distance it will travel from the ground to some given height (H) and record this value.
- With all in place, now Use a stopwatch and measure the amount of time (t) for a given trial for lifting the mass.
- Perform this measurement 3 times and average the time (t_{ave}) which is used.
- From the measurements, first :
- Calculate the new rotational kinetic energy of the motor using the pulley system by turning the linear kinetic energy of the system into rotational kinetic energy.
- The linear kinetic energy is determined from the distance traveled by the bag of marbles divided by the amount of time it takes for them to travel.
- The gravitational potential energy of the system is determined by the mass of the bag of marbles times the acceleration due to gravity times the distance (height) traveled.
- Here the change of gravitational potential energy (PE_g) over the amount of time (t_{ave}) as a power along with an additional factor of the rotational kinetic energy treated as a power of the new disk system will be used in combination to determine the Power Output#4 in watts for our system.(see calculation section for this efficiency calculation)
-
- **Analyzing the Data in All Activities :**
-
- Note : Each of the Sections will first be analyzed for their Power Input and Power Outputs as needed – this will be followed by the Efficiency Calculations for each of the Activities respectively since these are independent and require the Output Work from the prior Activity (which now acts as the Input to the Activity in question) and the Output in a given step is the measurement from that Activity :
-
- **Efficiency Calculations :**
-
- Initial Efficiency of Activity 1 :
- Note that this cannot be done if using the Sun
- For all other calculations one merely takes the ratio of the Work Output, which is the energy of the light source over the Work Input to the light source (such as the battery power or wattage of the bulb)
-

- Efficiency of Activity 1-2 :
-
- This is the ratio of the Work Output as determined from the Power of the Solar Cells as determined from the calculation of Power from the measurements of Voltage and Current to the Work Input which is the Work Output from the last Activity which is the Power of the light source
-
- Note : Independent of which light source used, you have determined its power which acted as initial work out #1 and now it will be used as Work In #2.
- In this calculation, use the readings from the Solar Cell (Voltage and Current) and determine Power from this. This Power will be used as Work Out #2
- Now use Work Out #2 & Work In #2 to determine Efficiency for this step of the Activity
-
- Efficiency of Activity 2-3 :
-
- This is the ratio of Work Output, which is the Rotational Kinetic Energy of the Motor being treated as Work to the Work Input of the Solar Cells to the Motor
-
- Efficiency of Activity 3-4 :
-
- This is the ratio of the combination (addition) of the gravitational potential energy of the bag of marbles lifted plus the rotational kinetic energy of the pulley disks of the motor to the input energy of the motor which is the initial rotational kinetic energy of the motor as determined in Activity 3
-

Data :

Air Calorimeter for Activity 1B :

Dimension	Measure (cm)
L : Length	
W : Width	
H : Heigth	
V : Volume	cm^3

Mass of Air in Air Calorimeter (m_{air}) : _____ g
    Determined from air density and calorimeter volume

Activity 1A :

Flashlight :
Voltage (V_{1A}) : _____ V
Current (I_{1A}) : _____ A
Power (W_I) : _____ Watts
    Power from calculation and is Work Input#1

Distance of photometer : 1.0 m
Photometer Measure (L_{1A}) : _____ lux
Calculated Work Output #1 : _____ watts
    Note that this is Work Input #2
    This measurement is also for the Lamp if there is no Air Calorimeter

Activity 1B :

Noted Lamp Wattage (W_I) : _____ watts
    This can be Work Input #1

Calculated Work Input #1 (W_I) : _____ watts

Use of Air Calorimeter :

Time (min)	Temperature (°C)
0	T_i
2	
4	
10 – 16	T_f

Change of Temperature (ΔT) : _____ °C
Quantity of Heat added to Air Calorimeter : _____ J

Calculated Slope of Line (ΔTemp / Δtime) : _____

Calculated Power Loss due to Heat in Air Calorimeter : _____ watts

Net Work Output of Light (W_{O1}) : _____ watts
    Calculated from $P_{Net} = P_{max\ input} - P_{loss}$
    Note that this is Work Input #2

Efficiency for Activity 1, 2, 3, 4 :

Note : To change this to all other Activities, simply change the 1 to 2
       and so on as needed for each of the Activities (2, 3, 4)
       This means you need to do this calculation for each Activity where
       the proceeding Activity is the Input Work and the Work measured
       in that Activity is the Output Work

$$\textbf{Efficiency} = \frac{\textbf{Work Output \#1, then 2, et al}}{\textbf{Work Input \#1, then 2, et al}}$$

Efficiency = _____

Activity 1C :

Use Solar Constant or Do Solar Constant Activity
This is Work Output #1 and is Work Input #2

Activity 2 :

From Activity 1, we now have Work Input #2

Solar Cells :

Total Voltage (V_{SC}) : _____ V
Total Current (I_{SC}) : _____ A

Total Work Output#2 (W_{O2}) : _____ watts
        Acts as Work Input#3 (W_{I3})

Activity 3 :

Disk diameter (d) : _____ cm
Disk radius (r) :   _____ cm
Disk mass (m_d) :   _____ g

Angular Speed of Motor (ω) : _____ rev/s

Rotational Kinetic Energy of Motor (KE_R) : _____ J
        Note : This is the assumed Power - Work Output #3
        Note : This is the Work Input #4

Activity 4 :

Inner Disk :
Disk diameter (d) : _____ cm
Disk radius (r) :   _____ cm
Disk total mass (m_T) :   _____ g

Circumference of Disk (C) : _____ m

Length or Distance Traveled (H) : _____ m

Mass of Marbles used (M) : _____ kg

Potential Energy of System (PEg) : _____ J

Time to travel distance H (t) : _____ s

Calculated Linear Speed (v) : ($\frac{H}{t}$) : _____ m/s

Calculated Rotational Speed (ω) : _____ rev/s

Rotational Kinetic Energy (KE_R) : _____ J

Total Energy of System (E) : _____ J

Total Power of System (Work Output #4) : ($\frac{E}{t}$) : _____ watts

Calculations :

Be sure to use your Slide Rule!

Efficiency :

$$Eff = \frac{useful\ Energy\ Out}{Energy\ In}$$

$$\mathbf{Eff = \frac{W_{out}}{W_{in}}} \qquad (W = Work)$$

Power :

$$\mathbf{P = \frac{W}{t} = \frac{\Delta E}{t}} \quad (General\ Formula : W=Work, \Delta E = Change\ of\ Energy, $$
t=time)

$$\mathbf{P = V*I} \qquad (Electrical : V=voltage, I=Current)$$

$P = N*S$ (N = No. of Lux or L_{1A} and S is number of square meters)

$P_{Net} = P_{max\ input} - P_{loss}$ (Determines Net Power in Light Output)

Energy :

$W = F*d$ (General Formula for Work : W=Work, F=Force, d=distance)

$W = \Delta E$ (General Formula : W=Work, ΔE = Change of Energy)

$PE_g = m*g*h$ (Gravitational Potential Energy : m=mass, g=acceleration due to gravity, h=height)

$PE_{el} = \frac{1}{2}*k*x^2$ (Elastic Potential Energy : k=spring constant, x=displacement of spring)

$KE = \frac{1}{2}*m*v^2$ (Kinetic Energy : m=mass, v=velocity)

$KE_r = \frac{1}{2}*I*\omega^2$ (Rotational Kinetic Energy = Moment of Inertia * Angular Speed)

$Q = m*c*\Delta T$ (Heat Needed-Emitted = Mass * Specific Heat Capacity of a Material * Change of Temperature)

Other Relations Needed :

$\Delta N = N_f - N_i$ (The change of a given variable is the difference of a final value and an initial value)

$m = \frac{\Delta Y}{\Delta X}$ (Slope for a given line on a graph)

Area of a Square : $A = L*W$
Volume of a Rectangular Solid : $V = L*W*H$

$C = 2*\pi*r$ (Circumference from the radius of a circle)

$r = \frac{d}{2}$ (radius is half the diameter of a circle)

$F = m*a$ (Net Force = mass * net acceleration)

$F = k*x$ (Hooke's Law : Force = Spring Constant*Distance Stretched)

Speed :

$$v = \frac{\Delta d}{\Delta t}$$ (Linear Speed = Distance Traveled / Elapsed Time)

$$\omega = \frac{v}{r}$$ (Angular Speed = Linear Speed / Radial Distance)

Moment of Inertia Formulae :

$$I_{disc} = \frac{1}{2}*m*r^2$$ (m = mass, r = radius)

$$I_{rod\text{-}mid} = \frac{1}{12}*m*L^2$$ (L = length)

$$I_{rod\text{-}end} = \frac{1}{3}*m*L^2$$

Constants :

1 m = 100 cm
1 kg = 1000 g
1 lux = 1 lm/m^2
1 kg = 2.2 lbs
1 in = 2.54 cm
1 ft = 12 in
1 lb = 16 oz.
1 oz. = 28.3 g
1 cal = 4.184 J
1 Watt = 1 J/s
1 kW = 3,414 BTU/hr

Acceleration due to Gravity : g = 9.8 m/s^2 = 980 m/s^2 = 32.4 ft/s^2

Specific Heat Capacity of Water
$$c = 1.0\,\frac{cal}{g*^\circ C} = 4.186\,\frac{J}{g*^\circ C}$$
Specific Heat Capacity of Air (at room temp of 25° C)
$$c = 0.242\,\frac{cal}{g*^\circ K} = 1.012\,\frac{J}{g*^\circ K}$$

Actual Solar Constant : 1.366 kW/m^2
Corrected Amount received at Earth for Solar Constant :
342 W/m^2

1 kW = 10^3 watts
1 kW = 3,414 BTU/hr
1 BTU/hr = 2.93 x 10^{-4} kW

Constant Used to Convert Light Into Power
1 lux = 1.46 x 10^{-7} Watts/sq.cm. (at 555 nm (green light))

1 lux = 1 x 10^{-4} lumen/sq.cm.
1 lux = 0.0929 lum/sq.ft. (also ft-candles)

Conclusion :

One of the first set of questions to consider are these : What are the types of energy involved in your system? What type of energy transformations take place in this system? When it comes to efficiency, what becomes of the energy that seems to be 'lost' when going from one step to the next? How do you account for it (assuming conservation of energy applies)?

What do your results show for Efficiency for component parts? How do you think this relates to the whole system? Why do you think some subsystems were more efficient than others? Would do you think can be done (and perhaps tested) to change or effect efficiency of the whole system or some subcomponent of it? If the system were less complicated, would efficiency be a larger value or not (bearing in mind, would it depend on the components involved)?

Alternative Tests to consider for comparison (with parental permission and supervision) :

1) Try a different wattage bulb
2) Try a different type of bulb (fluorescent)
3) Try a different distance for a given bulb intensity
4) Try a different size solar cell, or more solar cells
5) Try a different engine size
6) Try a different number of marbles as the load to be lifted
7) Try a different initial energy input of your design and testing (with parental permission and supervision)

Activity # 20
Graphically investigating the Inverse-Square Law of Light
Grade Level : High School
Math Level : Challenging

Inverse-square Law and Light Activity —

The heart of all of science is to not just notice phenomena, but the connections that a given variable has with other related variables. One of the most universal relations is known as the **Inverse-Square Law**.

In Physics, as well as all of the sciences, any physical law that states that some given physical quantity or intensity is inversely proportional to the square of the distance from the source of that physical quantity is said to be an **inverse-square law**.

The Inverse-Square Law basically applies when a given force, energy, or other conserved quantity is radiated outward in a radial manner from a source so that the surface area of this radiated area is spherical ($4*\pi*r^2$) hence proportional to the square of the radius, the force, energy or other quantity must spread out over this area. so this quantity must too diminish in intensity in an inverse proportional manner with the distance squared. It turns out that there are a number of phenomena that follow this general relation.

This activity explores the illumination of light, which all of us notice naturally. When we are close to a light it is clearly much brighter than when we are at a distance from it. **(Note do not stare into active light sources as this can damage ones eyes)**.

One of the first noted and mathematically explained **inverse-square relations** was the invisible force of gravity itself noted by Sir Isaac Newton in 1687. In his work, the Principia, he not only outlines his concepts of motion but discusses mathematically the force of gravity. He notes that the force of attraction of gravity falls off with the inverse-square of the distance between the centers of masses, such as the force between the planets and the Sun.

For example, this means that if there are 2 equally massed planets, yet one is twice as far from the Sun as the other, it would have only ¼ the gravitational force acting on it as compared to the other ($1 / 2^2 = ¼$). For the same mass at 3x the distance the force of gravity acting on it as compared to the closest one is only 1/9 as strong.

If on the other hand, let's consider 2 equal masses and move one mass towards the other, the amount of force rapidly increases. At half the original distance ($1/(1/2)^2$), the force of gravitational attraction between them is 4 times what

214

the original force was! Recall that a net force necessitates acceleration (Newton's 2nd Law) — all the more reason planets closer to the Sun move faster than those farther away!

Newton's Universal Law of Gravitation :

$$F = G * \frac{m_1 * m_2}{d^2}$$

Considering a Force with all other factors other than distance taken out :

$$\mathbf{F} \sim \frac{1}{d^2}$$

What is important about this relation in terms of mathematical analysis is that the product of the inverse-square of the distance and a given quantity (here force) will be a constant. We will make use of this in the Activity in terms of Light Intensity and the inverse-square of the Distances we use.

Later in history from Newton's time, August Coulomb found in 1785 that the attractive and repulsive force between unlike and like charges behaved much like gravity with regards to distance. That is to say, the amount of electrostatic force varies as the inverse-square of the distance between the charged objects.

Notice how fundamental forces like Gravity and the Electrical Force (2 of the 4 primary forces of the Universe) act in this manner. Imagine the surprise, excitement and wonder of the scientists who uncovered such things in those times. Nature was measureable, it was knowable, it made sense and seemed to have a simplicity and harmony to it.

It turns out that the Intensity of Light (which includes all forms of Electromagnetic Radiation) has this same relation between intensity and distance from the radiating source! With distance, the intensity of light falls off as the square of the distance.

Our goal is to examine the Inverse-Square Law and its application to the Intensity of Light. In the Activity, we use a Solar Cell (photovoltaic cell) connected to a Multi-meter that is at a measured distance from a luminous source (an incandescent lamp with a light that is on). We measure the Current (I) for the Solar Cell at the given distance (d) when illuminated by a light source. (Note : We are not using Voltage, since it will remain fairly constant with distances that are similar to each other as in our experiment (test this for yourself)).

It turns out that there is a relation between the square of the current and the power of the light. How this relation comes about is this way : Power for an electrical system can be determined as voltage time current (P = V*I) and from Ohm's Law (V = I*R) we can substitute for V the terms I*R and obtain (P = I^2*R). The resistance (R) here is treated as internal and constant to the solar cell (photovoltaic) and factored out, so $P \sim I^2$.

In the case of light sources, Illumination (B) is the number of lumens per meter squared ($B = \frac{L}{4*n*D^2}$)(the units of Illumination are lux). The Luminance (L) aka Luminous Flux is proportional to the Power of the Light Source ($L \sim P$) hence $L \sim I^2$. Though we measure light sources in terms of Power (Watts) they can be considered in terms of their luminance. In fact for a given wattage of a incandescent bulb, they average about 15 lumens per Watt while fluorescent bulbs (depending on power rating) range between 50-100 lumens per Watt.

Our Activity will place a simple incandescent light source at a distance from a photovoltaic cell connected to a multi-meter set up to measure the current (in milliamps) at varying distances from the light source. Both the Distance (D) and the current (I) are recorded. This data is graphed, mathematically recalculated, graphed again in two ways to determine the relation between light intensity (taken as current) and distance from the source to find the inverse-square law relation!

It is useful to examine this phenomena since it involves the idea of inverse-square relations, graphical and mathematical analysis, plus in the case of light, it is practical since when coupled with other interactions of light with matter (particularly reflection, absorption & re-emission, as well as refraction) it helps in understanding light and can be employed for light intensity needs for indoor as well as outdoor lighting considerations.

Purpose : To investigate through graphical and numerical analysis from measurements using a photovoltaic cell and a multi-meter the relation of light intensity and distance from the source.

Materials :

- Short Lamp with bulb (60 watt) (take off shade),
- Measuring tape (metric measures are best),
- Multi-meter ,
- Photovoltaic solar cell,
- Paper Towel tube or Toilet Roll tube cut to 5 cm cylinder,
- Sheet of dark Construction Paper,
- Graph paper,
- Long table (kitchen will do),
- Inverted Crate or Stack of books (to set multi-meter and solar cell on),
- Darkened Room,
- Small LED light to allow for writing, reading of information,
- Clear Tape,
- Slide Rule

Procedure :

1) Note : have the room as dark as possible, but have a small LED light to be able to do work as needed.
2) First cut the paper tube (paper towel tube or toilet roll tube) to a length of not more than 5 cm.
3) Attach to one end the dark construction paper and cut out the hole.
4) This tube with a shield will be placed in front of the solar cell so that only a constant area with some protection will act to only allow in the light along that path and keep out all other stray light sources.
5) With lights in the room on, set the lamp at the far end of a table.
6) Unfurl the measuring tape from the base of the lamp away from it across the table.
7) It is best to use an inverted crate (or stack of books if the crate is not available) to place the multi-meter and solar cell on.
8) The Solar Cell must be set up so as to be at the same height of the lamp and facing it (the tape can help hold it up) (books, magazines and the like can be used for either the lamp or the crate to establish a level environment so that the solar cell and light bulb are on the same level).
9) Set the solar cell so that it is the prescribed distance as noted in the table for data.
10) Note that the tube assembly created in steps 2,3,4 is placed in front of the solar cell.
11) Attach the multi-meter so that it will read current (I) measured in milliamps.
12) Turn on the bulb and turn out the other lights in the room.
13) Note : Do NOT stare directly into the light when taking measurements.

14) For each reading, always let the system sit and stabilize for a few seconds. The values may oscillate around a couple of numbers – take the one that seems to be the most frequent.
15) Once you are done with all of the measurements, turn on the room lights and turn out the lamp. Disconnect the multi-meter and solar cell.
16) Follow the remainder of the directions in the Calculations portion.

Data :

Measuring the Inverse-Square Law Indirectly
(through Multi-meter Current Readings & Power Calculations)

Distance (cm)	Current [I] (mA)	Current^2 [I^2]	$\dfrac{1}{D^2}$
10			
20			
30			
40			
50			
60			
80			
100			

Calculations :

Be sure to use a Slide Rule for all of your calculations!

1) Use the Slide Rule and find the squared value of the current (I^2) and place this in the appropriate column. (This uses readings from the D Scale and looking at the A Scale).
2) Note that Electrical Power is proportional to the square of the current in the circuit. ($P \sim I^2$).
3) First use the D Scale looking up the distances used in the Activity and find the corresponding value on the C1 Scale (its inverse). Be sure to watch the exponents here! Fill in this portion of the Data Table.
4) Create a Table of log(distance) and the log (current^2) values as read from the Slide Rule. (Read the data from the D Scale and find the log values on the L Scale).
5) Graph on the Y-axis Current^2 (which is related to Power which is related to Light Intensity) vs. Distance on the X-axis. This curve should approximate an inverse-square relation.
6) To test the idea of the relation in the Activity :
7) Calculate the Product of Current^2 and $\dfrac{1}{Distance^2}$ values. If done correctly, they should be approximately the same. Add these values up and divide by the number of data points to determine an Average value.

8) Graph I^2 (Y-axis) vs. $\frac{1}{D^2}$ (X-axis).

9) Draw a best fit line and determine slope for this graph. It should be approximately the same as the average of the product!

10) Graph on the Y-axis the log(current^2) and on the X-axis log(distance).

11) Through the log-log graph data, draw a best fit line and calculate the slope of this line (which should be near -2).

12) Your relation will reveal the connection between the variables. If it is at -2, this means that the light intensity (which is measured by current^2) is inversely related (hence the negative sign) to the square of the distance (light intensity $\sim \frac{1}{\text{distance}^2}$)

13) Though the Slide Rule is a recommended tool, all of the calculations can be done with a graphing scientific calculator or the use of a spreadsheet program. In these calculations you have to generate a table of data, graph it, and then find the slope and/or equation of the best fit line for the data. Other formula calculations can be done with these tools as well.

Formulae :

P = V*I = I^2/R
   (Note 'R' here is assumed to be constant and factored out –
   we are noting that P $\sim I^2$ and that P $\sim$ Light Intensity)

Product = $I^2 * \frac{1}{D^2}$

Average = $\frac{\Sigma(\text{products})}{\text{number of products}}$

slope m $= \frac{\Delta Y}{\Delta X}$

m = $\frac{\Delta(I^2)}{\Delta(\frac{1}{D^2})}$

m = $\frac{\Delta \log(I^2)}{\Delta \log(D)}$

Conclusion :

Examining the data and the graphs one should be able to find the inverse-square law relation of light intensity and distance. Be sure to take into account all sources of error and redo this as necessary.

Activity #21
Inverse-square Law for Sound Activity –
Grade Level : High School
Math Level : Challenging

The heart of all of science is to not just notice phenomena, but the connections that a given variable has with other related variables. One of the most universal relations is known as the **Inverse-Square Law**.

In Physics, as well as all of the sciences, any physical law that states that some given physical quantity or intensity is inversely proportional to the square of the distance from the source of that physical quantity is said to be an **inverse-square law**.

The Inverse-Square Law basically applies when a given force, energy, or other conserved quantity is radiated outward in a radial manner from a source so that the surface area of this radiated area is spherical ($4*\pi*r^2$) hence proportional to the square of the radius, the force, energy or other quantity must spread out over this area. so this quantity must too diminish in intensity in an inverse proportional manner with the distance squared. It turns out that there are a number of phenomena that follow this general relation. The more extensive writing on this idea is found in the first Activity on Inverse-square Law for Light (#20).

This activity explores sound, which all of us notice naturally (in a frequency range of 20 Hz to 20,000 Hz). When we are close to a source of sound, it is clearly much more intense than when we are at a distance from it.

It turns out that the Intensity of Sound (which includes all wavelengths of it) has this same relation between intensity and distance from the Sound source! With distance, the intensity of Sound falls off as the square of the distance.

What this means is this : Let's say we have an agreed upon measurement of sound intensity (I) at a given distance (d). If we then double the distance and use our measuring device again, we will find for the same source of sound (now twice as far away) it will have an intensity of ¼ (which is $1/2^2$) of the original intensity.

Sound Waves are mechanical waves and like all waves possess energy. Being mechanical waves means that they require a medium to travel. We normally refer to sound travelling through the air, but it can travel through liquids and solids as well. In the latter case of solids and liquids we experience, measure, and see the effects of these waves as earthquakes. Sound waves are clearly much less intense in energy, but measurable none-the-less.

Most waves are measured by their energy transport as energy that is passing a point. We use the term Intensity to describe this. Since we are measuring energy per unit area and they are moving so it is also per unit time. But energy per unit time is Power (joules/sec) so sound intensity is really power per unit area and the unit of area is the meter-squared, so the basic unit is joules per meter-squared and recall that a joule/sec is a watt, so we can also say watts/meter-squared.

To measure sound intensity we use the unit decibels but and it does not use these units of watts per meter-squared, however. Decibel levels are taken relative to a standard, such as the level at which we can hear so is expressed as a ratio of the intensity in question, L_2, and the base intensity L_1 which is the threshold value. This means that decibels are actually unit-less numbers.

$$\text{Decibel Level} = 10*\log(\tfrac{L_2}{L_1})$$

Notice that the equation is not linear, but instead logarithmic. This is because human hearing operates in this manner – we can hear rather low intensity sounds and quite intense sounds.

Because sound is a wave it has the characteristic of spreading in a inverse-square pattern (see those other Activities and some of the above to understand this better as needed) the formula will come out and be rearranged this way :

$$L_2 = L_1 - [\ 20*\log(\tfrac{d_1}{d_2})\ ]$$

Our goal is to examine the Inverse-Square Law and its application to the Intensity of Sound. We will make a sound and measure it at various distances with a device that measures decibels. We will use the above equation to see how close we come to an inverse-square law relation.

Purpose : To take sound measurement data in terms of distance and sound level intensity (decibel readings) and to employ a known mathematical relation to see whether or not it follows the expected inverse-square law relation.

Materials :

- Sound Measuring device – this can be one of the following : 1) Multimeter with a sound meter (dB), 2) stand alone sound meter, 3) microphone connected to a stereo system with a decibel sound meter,
- Graph Paper,
- Measuring Tape,
- The room being used is best if it is tiled or cement floors (to prevent sound dilution),
- Consistent sound item : Can be one of the following : 1) a toy that has a clicking sound, 2) tapping a metal pan lid with a spoon as consistently as possible, 3) snapping one's fingers as consistently as possible, 4) other?,
- Slide Rule

Procedure :

1. It is always best to read through a Procedure, have an idea for the plan, gather the needed materials, and then work the plan.
2. Depending on the arrangement of the room used (best to have tiled, cement, or wood floors) choose the longest path for the measuring tape to be placed on the floor. It is best to have the sound meter in one corner while the sound source will be moving from chosen point to point on the measured out path.
3. Note the previous direction where the points of distance between the sound meter and the sound source are already chosen. You can use the table provided or develop one of your own.
4. Realize one possible problem with the experiment – as distance increases between the sound meter and the sound source, it would be nearly impossible for one person to do this – so it may be best to have a partner who does one or the other.
5. Another important idea is to not only do one reading per distance. It is best to do at least 3. Not all of the readings will be used in calculations – instead the average value for a given distance is to be used.
6. Be sure to test the system to see if operating the way it should and adjust as needed. One of the important things to note is that with the sound meter on, there is probably a continuous background noise (hence it does not read zero). If the system allows a zeroing out, employ it – if not – recognize this separate to the data table as what the regular reading is. Note : It does not have to be subtracted out, since it is part of each reading it essentially washes itself out of consideration.

7. At each given distance [d] perform the sound creation and record the sound level measurement [L]. Do this at least 3 times per distance.
8. When done, continue with calculations – or perhaps redo the experiment to see if the results are consistent. Another alternative is to try a different sound source at the same distances – note that this has its own table.
9. Before performing calculations, it may be a good idea to graph the data as it is to see what sort of shape it creates. Put the Sound Level Reading on the y-axis and the Distance on the x-axis and draw a line through the curve – it should appear as a inverse-square law relation curve (see those Activities for more information).
10. Calculations :
11. First determine an average value for each of the sound level measurements. These are used in the calculations.
12. Note : So as to not have to deal with sign conventions in the needed formula provided : first note that [ ] are absolute value bars
13. and then choose d1 < d2 or conversely L2 < L1
14. (hence L2-L1 will be negative and regardless of the outcome of the variables and the operations on them inside the absolute value bars, it will be negative in magnitude due to the negation outside them).
15. Go through your table and do each of the needed calculations using your slide rule. In each calculation the first data point used is first on the list followed by the one that occurs next (at a greater distance).
16. Given the precision of your tools used, you may have to use no more than 2 significant figures.

Data :

Distance [d] (m)	Sound Level Reading [L] (dB)
0.5	
1.0	
1.5	
2.0	
2.5	
3.0	
3.5	
4.0	

Calculations :

Be sure to use a Slide Rule for all of your calculations!

Formulae :

Determining Average :

$$X_{ave} = \frac{\sum_{i=1}^{n} x_i}{n}$$

Formula to Evaluate for Inverse-Square Law relation in Sound :

$$L_2 = L_1 - [\ 20*log(\frac{d_1}{d_2})\ ]$$

Note : So as to not have to deal with sign conventions first note that [ ] are absolute value bars and then choose d1 < d2 or conversely L2 < L1 (hence L2-L1 will be negative and regardless of the outcome of the variables and the operations on them inside the absolute value bars, it will be negative in magnitude due to the negation outside them).

Conclusion :

Examining the data and the calculations, plus a visual look at the graph one should be able to find the inverse-square law relation of sound intensity and distance. Be sure to take into account all sources of error and redo this as necessary. One of the ways to know that this is on track, with each doubling of distance there should be a decrease in decibel level of 6.

Activity #22
Determining the Wavelength of Light in a Laser
Grade Level : High School
Math Level : Challenging

Determining the Wavelength of Light in a Laser

Waves, in general, are phenomena in nature that transport energy from one place to another. Sound waves are compression or longitudinal mechanical waves that are vibrations of the medium they travel through (solids, liquids, or gases). Electromagnetic Waves are not mechanical and hence do not need a medium to travel through. All electromagnetic forms, like light, travel at the speed of light which is approximately 3×10^8 m/s. [More description depth of Waves is found in the Slinky Wave Speed Activity]

All waves also exhibit other unique behaviors in the presence of matter, such as reflection, refraction, and diffraction. Some Activities examine some of these (Index of Refraction which also discusses Reflection) . In this Activity, the behavior of a wave passing through a narrow opening, commonly called diffraction. Defined **Diffraction** is the bending of a wave around a barrier, like the edges of a barrier. This is a behavior that helped Christian Huygens win the day over Newton's idea of how to characterize light. Newton had proposed a model in which light was composed of individual particles. If it were as it passed through a narrow opening, the most that would happen is that the majority would continue in a straight line, while some would create a fainter and fainter spray pattern from the center.

What was found, though, when light was passed through a single narrow opening, and in a later experiment using a pair of slits, was that there was a central bright region, called **the primary maxima**, *but on either side of it* were **bright spots separated by dark gaps**. The first pair either side of the primary maxima are called the 1st maxima, the next are the 2nd maxima and so on in terms of names for them. Huygens' Wave model of light could explain this phenomenon through **constructive and destructive interference of the waves as it passes through the narrow openings**.

The best way to describe why this occurs is to use **interference of waves. Waves can interfere in a constructive or destructive manner**. In Constructive Interference, this is the addition of two or more waves since their amplitudes are in the same direction. These result in the bright spots. In the case of Destructive Interference, two or more waves have amplitudes in different directions hence subtract or diminish the outcome. This corresponds to the dark region between the bright spots.

Instead of using a single or double slit arrangement, this activity uses what is called a diffraction grating which can have several hundred lines per millimeter! You have to know the number of lines per millimeter to successfully do this activity since its number will affect the pattern produced. A good thing to do is to have two different gratings and try the activity twice to see the outcome.

When using a diffraction grating, the large number of slits results in the same sort of diffraction pattern (dependent on the number of slits) that occurs with one or two slits. The geometry of this situation allows one to find the wavelength of the light itself! This is because the distance traveled by one beam through one slit corresponds to a distance equal to a multiple of the wavelength of the light from an adjacent slit. They are in what is called phase with each other. These waves will add up on the wall constructively where they hit and produce the bright spot.

This Activity takes advantage of the diffraction behavior of waves and their resulting interference in order to determine from geometry and careful measurement the wavelengths of light, such as that from a laser.

A quick question must be asked and addressed which is : Is laser light any different than ordinary light? Laser light is actually a terms that is an acronym. It stands for **Light Amplification by Stimulated Emission of Radiation**. It is a means for emitting electromagnetic radiation (like any ordinary light does) with a little more focus as compared to ordinary light. In essence, it is concentrated light. First, laser light has waves that are in step with each other, this is called coherence. Light from a bulb is not this way and given off in a random manner, hence might be called incoherent light. This coherent light is composed of waves of identical frequency, phase, wavelength, and polarization. Though this may seem different, it is not. We are using the best source of light so that we have consistency in the frequency and wavelength plus it is a naturally narrow beam. Caution must be practiced however so do not look into it or even let it bounce off and strike you in the eyes from highly reflective surfaces.

Purpose : To use the property of diffraction of light and geometric optics analysis to determine the wavelength of a common laser.

Safety Notes : Have Parent permission and supervision for this activity. Do not shine laser light into anyone's eyes. Always exercise caution when using a laser.

Materials :

- Laser (HeNe, Green, or other),
- Meter Stick,
- Diffraction Gratings (of known slit width, marked lines/mm),
- Measuring Tape,
- Table,
- Tape,
- 2 Pieces of cardboard or 2 toilet paper towel tube,
- Roll of Paper or taped together sheets to act as screen which will have marks made on it (dull finish to paper is best),
- Slide Rule

Laser Information & Other Pre-Preparations :

1) Can use regular small lasers. The key is that it must be placed in a v-shaped box parallel to the table (see step 3).
2) Wavelengths of various common lasers that may be used : Red HeNe Laser : 633 nm, Green : 543.5 nm, Orange : 612 nm, and yellow : 594 nm (1 nm = 1×10^{-9} m)
3) For the v-shaped holder : Either use a thin piece of cardboard folded in an accordion fashion so that it creates a v-pocket for the laser OR split a cylindrical toilet paper towel tube vertically and use each half circle as sides so that looking down on it there is a channel that is v-shaped.
4) Create a 2nd v-shaped form which the diffraction grating can set in. Note that when activated the laser has to hit and go through the grating, so work with the cardboard pieces or tubes so that this happens. A simple solution is to place the laser v-shaped platform on a thin book so as to elevate it slightly.
5) Diffraction gratings can be found on line easily at low cost. Average ones are $2-$3 each and have 13,500 lines per inch (1 in. = 2.54 cm) – Note : Depending on cost, using 2 different gratings in 2 separate trials (i.e. different lines per mm) is a good idea for comparative purposes. Record this value (q) on your data table.
6) Place the Table adjacent to a wall long-wise. On the wall tape the paper roll to act as a screen across greater than the width of the table (over 1 m and up to 2 m in length). (Best to use dull finish paper).
7) Note : Test the system so that a pattern is seen and works well. There is no need for further adjustments once the pieces are all in place and operational.
8) **SAFETY at all times – Must have parent permission and supervision for doing this activity. Lasers are not to be looked at or directed into anyone's eyes. Also be cautionary with laser light bouncing off metallic or shiny surfaces.**

Procedure :

1) Read through and conduct the pre-preparations so that the laser goes through the diffraction grating and has various maxima on the screen. Be sure to have measured the distance from the diffraction grating to the central maxima (L).
2) Note : Do not measure with the laser on. The distance (L) is measured with the laser off.
3) It is best to mark where the central maxima is with a pencil on the paper screen and the central portion of the 2 left and 2 right maxima (you can do more if interested). Be careful in marking and do not look into the laser. Use of a dull finish paper is best so as to avoid reflection. Face only the paper when marking it and do not look back to the laser.
4) Turn off the laser and take down the paper screen for measurements.
5) Note : You are not taking measurements with the laser on. You only mark the paper where the maxima fall. Measurements are done after the laser is deactivated and the paper is taken down to be measured.
6) Once all markings are done fill in the data table from measurements taken. Though it is noted to the meter, realize that you are measuring with a tool that can measure to the 0.1 of a mm.
7) The Data your are recording here is the distance from the central point to each of the subsequent maxima (x).
8) Follow the directions in the calculation section to determine the wavelength of light for the laser and compute percent error.
9) Calculations :
10) Though the Slide Rule is a recommended tool, all of these calculations can be done with a regular or scientific calculator. Some scientific ones even have built-in averaging formulae. For those who like spreadsheets, the data can be typed in and the formulae then also be typed in its own cell where the formula references each of the measured variables in their respective cells, for example B1..BN has the measurements and values used in the equation while BN+1 has the formula for all of these variables (why not A? Simple – use it to label you variables)
11) Determine the distance between the lines on the diffraction grating (d). (Be sure to convert to m).
12) Determine the distance from the diffraction grating to each of the maxima (Lo) (which is the hypotenuse) using the Pythagorean Theorem (the other sides are L and a given x)
13) You can either determine the sine of the angle in question from your data are merely move ahead and use the wavelength formula (since the sine is now expressed as a ratio of the measured and calculated sides of the triangle – see diagram).
14) With a calculated wavelength (λ) compare it to the known values and compute percent error.
15) An alternative is to determine all of the wavelength values for the maxima used and then average the values as well.
16) If other diffraction gratings : Other Experiments : If you are using more than one diffraction grating , redo all the steps and have a new screen.

Photo for Set Up :

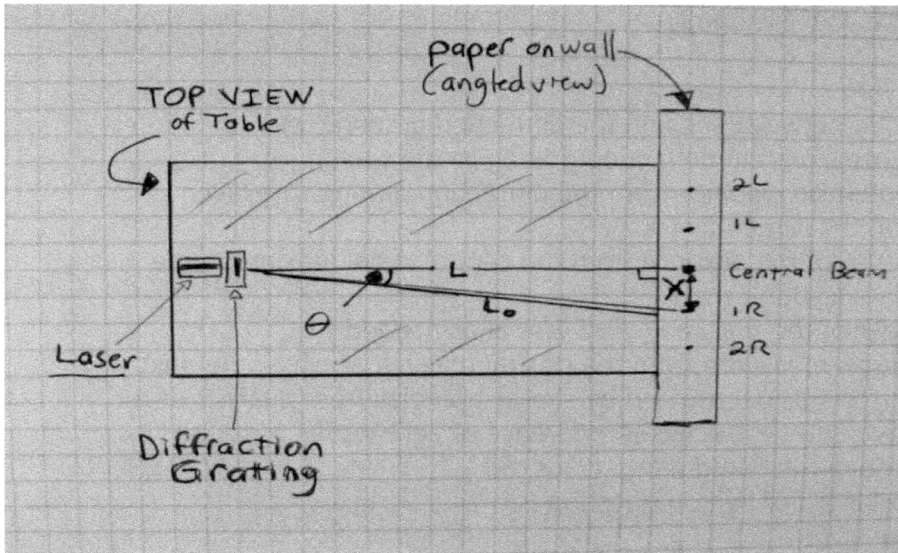

Data :

Number of lines per meter on the diffraction grating : _____ [q]
(Note – read the grating, conversion needed)

Spacing of Lines in Grating [d] : _____

Distance from Grating to Screen [L] : _____ m

n	x (m)
1 (L)	
1 (R)	
2 (L)	
2 (R)	

L : Left of center, R: right of center
Note that 'n' is 1 or 2 in the given case.

Calculations :

Be sure to use your Slide Rule!

1 m = 1000 mm
d : The spacing between lines on the diffraction grating (m)
L : The distance from the diffraction grating to the screen (m)
L_o : The hypotenuse distance from the diffraction grating to a given maxima on the screen (m)
x : The measure of the distance from the central initial maxima to the maxima in question (m)
n : Is the order of the maxima (n=1,2,et al as needed)
Θ : The angle as measured from the central maxima at which the next maxima occurs as determined by similar triangles from analysis of the constructive and destructive interference of light waves passing through the diffraction grating. ($^\circ$ or rad)
λ : The Wavelength of light in question (m)

Needed Formulae to Find the Wavelength :

$$d = –$$

$$L_o{}^2 = L^2 + x^2$$

$$Sin\,(\Theta) = \frac{x}{L_o}$$

$$n*\lambda = d*sin(\Theta)$$

Wavelength Formula (derived) :

$$\lambda = \frac{d}{n} * sin\,(\Theta) = \frac{d*x}{n*L_o}$$

Percent Error Formula :

$$\%E = \frac{[\text{Experimental Value-Accepted Value}]}{\text{Accepted Value}}*100\%$$

Average :
t : number of trials

$$\lambda_{ave} = \frac{\Sigma\lambda}{t}$$

Possible and Good Approximations Formulae in this Activity :

1) If Θ is used in radians, then $\Theta_R \sim \sin\Theta°$
   To use – find Θ from

   $$\tan\Theta = \frac{x}{L}$$

   $$\Theta_R = \frac{2*\pi}{360°}$$

2) $\sin\Theta \sim \tan\Theta$ for small angles
   This means for small angles, the ST scale on the slide rule can be used. What is being done here is we are using L instead of L_o and saying :

   $$\frac{x}{L_o} = \frac{x}{L}$$

Conclusion :

The best way to conduct the activity is to do it more than once and be careful and certain of your numbers. Know that there is a certain level of precision in your measurements, hence this affects the accuracy in your results. The best measure is to determine percent error in your wavelength determinations.

Activity #23
Using a Microwave to Determine the Speed of Light Activity
Grade Level : High School
Math Level : Calculating

Electromagnetic Waves are quite familiar to all of us. We see them in the form of Visible Light, which represents only a small portion of the electromagnetic spectrum. There are many others, however, we hear of often and are used very frequently by humans in their devices and daily lives. We use X-rays to examine the interiors of items, even ourselves when it comes to our bones. We hear of the harm and damage of Ultraviolet Rays and the need for sun block. Infrared rays are used for looking for people, animals, and the like at night. All of radio, TV, satellite phones and the like use a portion of the electromagnetic spectrum to send and receive messages and information which is called collectively Radio Waves. Also in there are Microwaves, not just the device but the waves the device uses to heat water molecules (to get them to vibrate) to heat food and cups of coffee and such.

Other than the fact that all of the waves mentioned are composed of the same parts – a magnetic wave and an electric field wave – there is one other unique feature all Electromagnetic Waves share in common which was found in Maxwell's Equations describing these waves and found to be a cornerstone of the Universe by Einstein and that is : In a vacuum all electromagnetic waves travel at what is classically called 'the Speed of Light'. Though this term is used, it is only because of human study of light first and the others later on. This value is listed in the Calculations section below and will be used in the Activity that follows and is normally seen as 3.0×10^8 m/s or 186,000 mps (that is miles per second not hour as we are normally expecting). If one could send a beam in a circular path around the world, it would circle the globe 7.5 times in one second!

The Microwave uses a device called a magnetron to generate microwaves from electrical energy usage. These waves are contained in the oven chamber even with the perforated screen at front because these holes are smaller than the wavelengths of the microwaves, so they cannot escape.

We might think that the waves in the microwave just randomly bounce around but they do not. The waves set up what is referred to as a Standing Wave Pattern. This is much like taking a slinky spring or a jump rope and making it move with just enough energy so that the item moves in a rhythmic and stable fashion, such as when two friends are spinning the jump rope just right to make what appears to be a half-wave from the side. It appears that the wave is rather stationary and not moving, hence its name. In the case of the jumping rope example, The two people are at what are called nodes and the high point in the middle is the wave crest. What if the two people with the jump rope (or the person moving the slinky spring back and forth) do so even faster. They can reach another stable speed where a node appears in the middle between them and now there is a complete wave cycle present as viewed from the side. All of this discussion of standing waves applies to microwaves as well. The waves in the microwave set up a standing pattern. If one had a very small instrument that could measure at different spots across the microwave one would find that microwave energy would rise to a peak, then decrease and peak yet again as one moves across the microwave chamber. The peaks are where the wave crests of the microwaves are while the areas of low to no energy are the nodes. This is why microwaves have a rotating table in them! This way something that is being heated will pass through these regions of peaks and valleys repeatedly. Without rotation there would only be hot spots on the food while the other areas remain rather cold.

We wish to take advantage of these peaks and valleys, so we must disable the rotation of the platter by taking it out as well as the post that the platter sits upon. Now when we go to heat something, which we will do in the Activity, there will be hot and cold spots. We will use something like chocolate or marshmallows (see list below in Activity) and only heat it a very short time to cause just the initial melting of the surface. What we are after is the distance between these hot spots! That distance corresponds to ½ of the wavelength of the microwave wave. We merely have to double this distance and we are half-way to determining the speed of light for these waves.

We have to use what is classically called the Wave Equation.

$$c = \lambda * f$$

It states that the speed of a wave is equal to the product of its wavelength and the wave's frequency. Normally the variable v is used instead of c since this equation is for all wave types (electromagnetic and mechanical alike) but when 'c' is used it is referring to light and all of its associated electromagnetic wave forms since they all travel at a constant speed all the time as Einstein predicted and found.

In our case of our Activity, We have the wavelength, but what of the frequency? This can be found on the manufacturer's sticker for the microwave (be sure to have parental support and help in this) found on the backside or underside of the device. Most microwaves have a frequency of 2450 MHz (2450×10^6 Hz) or one could simply go to the internet and type in the model name and number and with some research uncover the number needed.

As noted this is called the Speed of Light, as this applies to any and all Electromagnetic Waves in a vacuum. One could readily note that there is air in the microwave, and this will technically affect its speed, but not much at all – in fact if properly done your answer should be within a small percentage of error and somewhat reliable to one perhaps even two significant figures.

By the way be sure to follow all the directions below, have parental permission and help, plus use food items that one will consume – don't be wasteful. It is important to note on the side, however, that this Activity is one of the rare ones – in virtually no case does one ever consume anything in a lab exercise, but here the Activity is designed to be safe and allow for that outcome! Enjoy and Explore :)

Purpose : To determine the speed of light from distances between melting spots on food in a microwave due to heating by a microwave standing wave pattern and multipl ed by the frequency of those waves

Materials :
- Microwave,
- Microwave-safe baking or casserole dish,
- Food Item to Use : Choose one from : Chocolate bar(s), Chocolate morsels, bag of mini-marshmallows, cheese slices - (note for choice – must be sufficient to span if not cover the dish being used in a uniform layer – part of this depends on the food's use afterwards – good idea, don't waste good food),
- Ruler,
- Toothpicks,
- Goggles,
- Slide Rule

Important Notes :

Note 1 : Must have parental help and permission in this Activity. There are a number of steps which should involve parental help, such as disabling the rotating dish and finding the frequency of the microwave on the back or bottom of the machine (Note – you might find it on the internet with some research on your model), as well as properly handling heating of food items. The key is that you have a microwave where the center post of the rotating mechanism easily comes out – if not, then you cannot – most lift right out. Be sure to read all directions before doing the Activity.

Note 2 : Do not overheat the Food Item being used. The range of time does depend on microwave power and the item chosen, but typically ranges from a handful of seconds up to not longer than about 40 seconds. The key is watching the item while heating and once melting is detected, then stop the process. – Also do not touch it – it may be too hot and you are going to measure it, you don't want to ruin it.

Note 3 : Goggles are noted and should be worn as a good practice for science lab work, but also in the remote chance of the food item being too hot (which if done correctly it should not be) and it bubbles.

Note 4 : Food Choice is by preference and should be done in the knowledge that this food will be consumed or used in some sort of recipe as it is.

Note 5 : The microwave-safe dish is best to be a dish that has a wall so that a rigid ruler can be put across it and one can read looking down at the food beneath the ruler above it spanning the dish rim.

Note 6 : There are others who have written about this lab and have used whipped egg whites or very-buttered bread instead of the list noted above.

Set-Up :
- Be sure to follow all of the Notes above in your procedure
- Before using the microwave it is best to record the frequency of it found on the sticker label (typically on the back of it), f. Record this in your data table. If unable to find it, perhaps try the internet or simply use the suggested value in the table in Calculations as this is the most common value
- In preparing the microwave for use in the lab, you are taking out the turntable and its post (typically the post is plastic and lifts right out) – This is so a dish can sit in there and not rotate when the microwave is on
- Be sure to use a microwave-safe dish and it is best if it covers as much of the microwave floor as possible but also have a span that can be spanned by a ruler that can rest on its rim
- In terms of the ruler, it is best to use a solid one (that is not flexible)
- In choice of food item to use, use one you plan to make use of

Procedure :
1. Be sure to follow all of the Notes listed prior to the Procedure, follow safety guidelines, such as wearing goggles, and also have completed the Set-Up before continuing.
2. Whatever your choice of food item, it will be referred to simply as chocolate from this point on for reference purposes in these directions.
3. Using a microwave-safe baking dish place chocolate in the bottom (note it is best to use solid-type with no other items in it and that it is reasonably thick). It will cover as much of the area as possible, particularly spanning from one side to the other.
4. Place the dish in the microwave in the center, close the door and turn it on. The amount of time will depend on the food type, its thickness, and power setting chosen, but in virtually all cases will be less than 40s and normally only run in the neighborhood of 20s.
5. The key to the amount of time heating is not only that it is short in time, but will depend on you keeping a careful eye on the food item heating. At the first signs of melting in a couple of spots, it is time to stop it. Do not let it overheat, bubble or even liquefy to any appreciable amount.
6. Be sure not to touch the item or move it too much to disturb it since you must measure the distance between the spots.
7. The number of measurements will depend on the number of spots. Each set of two spots in sequence are to have their distance between them measured. This is referred to as a Trial in the Data section and denoted by the letter S.
8. So this means Trial 1 is for the distance between spots 1 & 2, Trial 2 is for the distance between spots 2 & 3, and so on.
9. The best way to measure is to place the ruler across the rim of the dish and read by looking directly down where the centimeter side is above the spots in question.
10. Realize you are not starting at '0' with each measurement. For example let's say spot 1 is at 4.3 cm while spot 2 is seen at 10.8 cm – how far are they apart. You need to find the difference of these numbers and this is S (for this example it is 6.5 cm).
11. To help in determining where the 'spot' is use the toothpicks and poke them into the center of the melted spot on the chocolate, if soft, and if not, merely touch down to the center-point of the spot in question when reading its value on the ruler. Note do not force it into the chocolate as this may affect the readings of not only this spot but others. – In essence the toothpicks are guide posts. – Use only if needed.
12.

13. Once all the measurements are done, look at these values – they should be similar to each other if done correctly. Determine the average of these values (X_{ave}) this is the value of half of the wavelength of the microwave in question.
14. Now double the average value and this is the wavelength of the microwave wave for our equation ($\lambda = 2*X_{ave}$) [Be sure to convert this to meters]
15. With the wavelength determined, we can determine the speed of light from the wave equation since we have either read the frequency from the manufacturer on the microwave or are using the standard value provided in the table below.
16. How did your results turn out? Use the percent error formula to determine how close you came expressed as a percentage. Be sure to only use two significant figures as this is the extent of your measurements as well.
17. Be sure to properly clean up and reset the microwave. It is more fun to use food that is being used in a fun recipe. Enjoy! :)

Data :

Frequency of Microwave [f] : _____ (Hz)

Trial	S (cm to nearest 0.1) Distance between melting spots
1	
2	
Average	

Calculations :

Be sure to Use Your Slide Rule! :)

Average Value :

$$X_{ave} = \frac{\Sigma S}{n}$$

(S is the distance between points, n is the number of trials)

Wavelength :

$$\lambda = 2*X_{ave}$$

Wave Speed : (Speed is Wavelength times frequency)

$$v = \lambda*f$$
$$c = \lambda*f$$

Percent Error :

$$\%E = \frac{[\text{ Actual-Experimental }]}{\text{Actual}} * 100\%$$

Constants & Conversions :

$1M = 10^6$
$1G = 10^9$
$1m = 10^2$ cm
$c = 3.0 \times 10^8$ m/s (value to use here)
   [$c = 2.99792458 \times 10^8$ m/s]
$f = 2450$ MHz (most microwaves are this)
   [use this value if unable to find or read yours]

Conclusion :

How did your results turn out (percent-error wise)? How similar were the melting points in the measured distances between them?

Activity #24
Make a Radio and Measure its Signal Activity
Grade Level : High School
Math Level : Challenging

What we refer to as Radio is shorthand for radiotelegraphy from the times when it was called 'wireless telegraphy' and called 'wireless' in Britain. The shortened name 'radio' came from the late 1800s in France from French physicist Edouard Branly and the word radioconductor and picked up by the United States Navy in 1912 and was used often in the military. The first commercial broadcasts began in the 1920s in the U.S. (the first AM radio news broadcast was on August 31, 1920 in Detroit, Michigan at the station 8MK which is still operating today as an all-news format station WWJ and owned by CBS network).

The history of the radio has many inventors, scientists, and engineers the world over and is rather complex. Though the chief contenders for fame claim are Tesla and Edison, there were many physicists from England and Russia as well. (For example, in 1878 David E. Hughes recognized that sparks could be heard in a receiver when he was experimenting with a carbon microphone which was demonstrated to the Royal Society as early as 1880). Initially Marconi had the patent awarded to him. Eventually, however, in a 1943 Supreme Court ruling the term radio was defined by this decision ("A radio communication system requires two tuned circuits each at the transmitter and the receiver, all four tuned to the same frequency") and awarded the first patent to Nikolas Tesla (from 1897) though it was after Thomas Edison's application for a patent in 1885 which was purchased by the Marconi Company in 1891. Tesla had created the foundation devices as early as 1893 and presented them publically at the Franklin Institute in Philadelphia while addressing the National Electric Light Association. While Marconi had demonstrated the basic format of the radio in 1895 and transmitted a signal over 1 mile. From this it was found that the transmission range is proportional to the square of the antenna height and is called "Marconi's Law".

Radio is the transmitting of electromagnetic signals the modulation (changing) of these electromagnetic waves through free space whose frequencies below that of visible light. Like all electromagnetic waves, radio waves travel at the speed of light (3×10^8 m/s). Like all electromagnetic waves this form of radiation travels by the process of oscillating electric and magnetic fields at right angles to each other and at right angles to their direction of travel. In radio waves, the information is carried by the process of systematically changing (the term used is modulating) some property of these waves, such as amplitude, frequency, phase, or pulse width. When these waves come in contact with an electrical conductor (such as in the antenna) the electrically oscillating fields induce an alternating current in the conductor. This can be detected and then transformed into sound or other signals that carry information (i.e. the radio).

All radio systems have these elements : Transmitter, Receiver

A Transmitter, which has a source of electrical energy which generates alternating current of an intended frequency of oscillation. The transmitter has a system designed to change (modulate) some property as noted above (amplitude, frequency, phase, etc) of the energy produced to impress a signal on it. The transmitter sends this modulated electrical energy signal into an tuned resonant antenna which converts the changing electrical current into an electromagnetic wave and emits it to free space.

The emitted electromagnetic waves can travel directly or have their path altered through the processes of reflection, refraction, or diffraction. The intensity of the waves falls off as the inverse-square of the distance from the transmission source (see Inverse-square Law of Light Activity) as well as some of the energy being absorbed by various materials in the environment.

The waves are then intercepted by a tuned receiving antenna. This antenna will capture come of the energy of the wave and changes it back into an oscillating electrical current. The Receiver then demodulates these currents. This is the conversion of the electrical current signals into a usable signal form by a detector sub-system. The receiver is tuned so that it responds preferentially to the desired signals (hence not locking onto other signals) (i.e. selecting the station on the radio).

The early radios relied only on the energy collected by the antenna to create signals for the operator. Later invented devices such as the vacuum tube and the later transistor makes it able to amplify these weak signals.

Radio systems are not merely portable players and devices in our cars for AM and FM radio stations, but also includes walkie-talkies, radio-controlled remote controlled devices, used in the control of space probes and vehicles, communication in space, navigation, RADAR (Radio Detection and Ranging), cellular phones, along with other broadcasting such as TV which uses AM and FM signals.

What is AM and FM radio?

Both AM and FM are radio signals. AM stands for Amplitude Modulation where the amplitude of the transmitted radio signal is made to be proportional to the sound amplitude that is captured by the microphone, while the transmitting frequency remains unchanged. AM radio signals are affected by static and interference. Natural phenomena, like lightning, have radio emissions at the same frequency as these signals. AM radio stations broadcast with power levels as high as 500 kW originally (today in the U.S. and Canada are limited to 50 kW) and due to reflection off the ionized part of the atmosphere can be literally picked up world-wide.

FM radio signals, stands for Frequency Modulation, where there is amplitude variation at the microphone which causes the transmitter frequency to fluctuate. FM signals do not have the static and interference that often occur with AM signals since it is in a higher frequency area (VHF 30 MHz to 300 MHz) and have transmission ranges of 50 to 100 miles. Also they can penetrate buildings more readily than AM signals. At present, government, police, and fire services use narrowband FM frequencies.

The general idea of the simple Crystal AM radio :

Crystal Radio sets are AM radios and have been around for over half of a century and are essentially the essence of a radio in its simplest form. It has no power source, such as batteries, and only requires a few parts. The power source turns out to be the radio waves themseves that create electrical potential differences   (voltage) and since associated with conductors (wires composed of electrically conducting materials) current can flow.

The basic parts are : Earphone, Ground Wire, Antenna Wire, Inductor (coil of wire), Germanium Diode, and a Variable Capacitor (here in our Activity, Aluminum Foil – the variable capacitor is used to tune in a given frequency (i.e. the station)).

The Antenna is the material (here a wire not grounded) is a conductor that receives the radio waves and converts them into an electrical current. The electrical current flows into the radio circuit between the coil of wire (inductor) and the variable capacitor (aluminum foil). The inductor changes the speed of the electrons by slowing them down. The inductor is a coil of wire, the more coils the longer the path for the electrons, hence the greater the amount of time for the current to change. At the same time it is in parallel to the variable capacitor has a given value of electrical capacity. This creates a back and forth exchange between the coil and the capacitor so that the signal resonates at a given frequency. This frequency is associated with the radio wave frequency for the given station. This signal now flows through the diode which allows only the peaks in the current to flow through and allows it to flow only in one direction to the earphone (whose other wire is connected to the ground wire to complete the circuit) where there is an electromagnet (see the Electromagnet Activity) so that the current flows in a changing manner through a coil of wire wrapped around a magnet. The current affects the strength of the electromagnet which will attract the diaphragm of the earpiece (which is imbedded with iron   particles) so that it vibrates to produce sound.

The Activity is to construct just such a radio and measure the Power of the signal received with a multimeter (voltage and current readings).

Note : Do not do this Activity without parental permission and supervision. Do not use wires, electrical devices, and the like in an unsafe manner. This Activity does not involve the use of any electrical outlets, power sources, and the like. One of the critical aspects of this project is that one must have reasonably close and strong AM radio signals for it to be effective. Also the weaker the signal, the longer the antenna needs to be. There are other parts that can be used and with some research, one can find on the internet other configurations and materials to use for the crystal radio set. For example, you can create a variable inductor instead of a variable capacitor. There are even kits already prepared for construction and use.

Purpose : To construct a basic AM Radio from common materials, receive local radio station signals, and measure the power of the signal received through readings of voltage and current on a multimeter.

Materials :

- Aluminum Foil,
- Magnet Wire (22 or 26 gauge & length),
- Wire (20 gauge, 10 m long, will be cut – antenna & ground),
- Wires with alligator clip ends (to make assembly easy),
- Poster Board (can act as base for radio),
- Wire Stripper (if needed),
- Sandpaper,
- Multimeter,
- Earphone,
- Diode (1N34A) (Radio Shack pt no. 276-1123),
- Large Plastic Cups that can nest in each other (20, 24 or 32 oz),
- Masking Tape and Clear Tape,
- Note : Need to live an area where AM radio signals are available,
- Slide Rule

Note : The larger the cups used, the better. They can also be Styrofoam or Paper. Also the construction may involve the creation of several of these (so as to have different capacitors).

Note : Measuring Voltage and Current requires parental permission and supervision. Also these electrical measurements are only done with this type of radio and no other form! As with all electrical activities, exercise caution in using the tools properly and safely.

Set Up :

- The longer the Antenna wire, the better. One could use a wire that they connect to an antenna (car antenna, or an old TV antenna, hence the wire need not be as long. If the wire is the antenna, then it needs to have some length and be separated from the ground).
- Use the magnet wire and wrap it at least 100 times (or more as desired) around the paper towel tube and leave long enough ends so that it can be attached to the circuit to make the radio.
- Use sandpaper to remove the coating from the magnet wire ends.
- Cover half of the cylinder nearly completely around of one of the plastic cups. Tape with clear tape the foil in place. Make sure it is smooth and flat to the surface.
- Cover the other cup cylinder surface with aluminum foil as well ½ way around so that the same amount of area is covered.

- Use Clear Tape and tape to each of the cups a wire where the ends have been stripped. One wire per cup. Nest one cup in the other. The bare ends of the wire are out of the cup.
- Use alligator clip wires and attach one of the magnet wires to one of the two cup wires to the clip of the wire in use.
- Connect a second alligator wire to the other wire from the other cup and the other end of the magnet wire.
- The above description means that the Coil (the magnet wire tube) and the Variable Capacitor (the Aluminum Foil covered Cups) are in parallel to each other (see photos of items and diagrams below).
- To one of the free alligator clip wire ends attach the diode. (We will call this end 'A' since one more attachment will be made to it!)
- To the diode, attach one of the wires of the earphone.
- Attach the other earphone wire to the other free alligator clip wire end.   (We will call this end 'G' since one more attachment will be made to it!)
- To end 'A' attach the wire (or item) that is to become the Antenna.
- To the end 'G' attach a wire to become the ground, which can be a cold water faucet. Note : In either case of the Antenna or Ground they are not attached to electrical outlets or devices and you have supervision for such actions.
- The Radio should be operational now, but realize you may have to adjust the cups (your variable capacitor) and move the antenna plus check your connections. With patience and effort, you should be able to obtain a radio signal (provided you live within range of a strong enough signal).
- Now move on to the measurements in the Procedure.

Photos from Set Up :

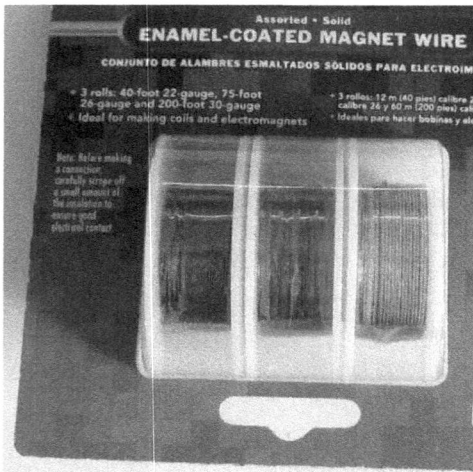

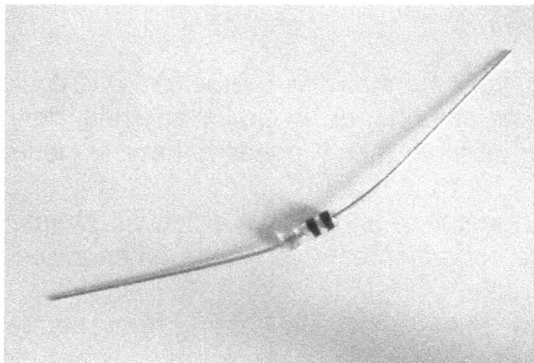

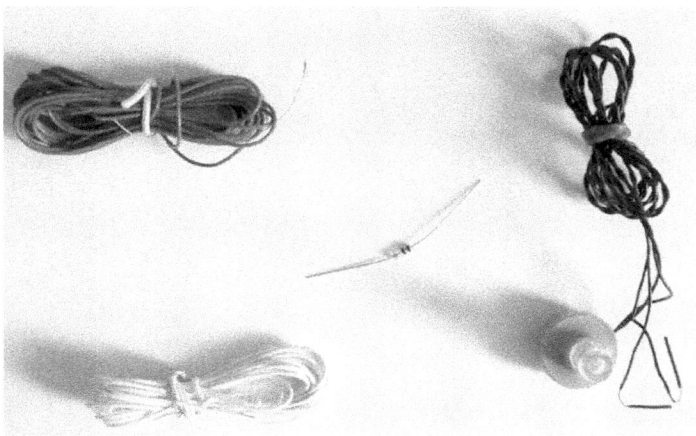

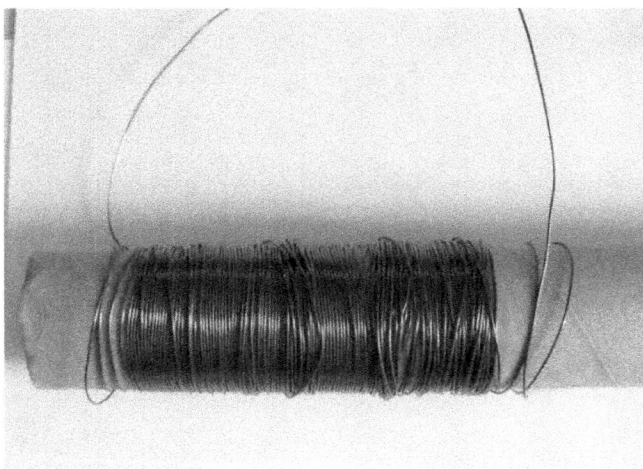

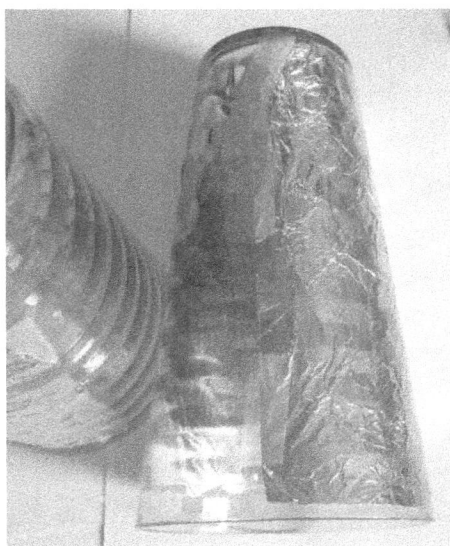

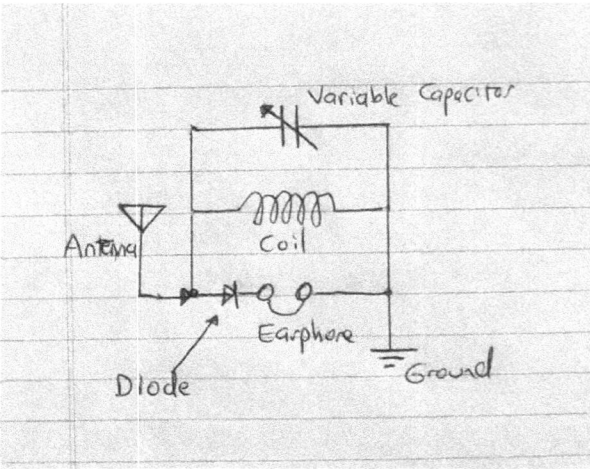

Variable Capacitor

Antenna

Coil

Diode

Earphone

Ground

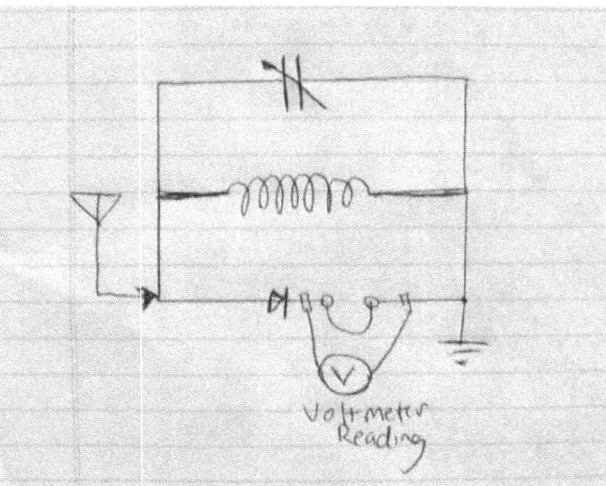

Voltmeter Reading

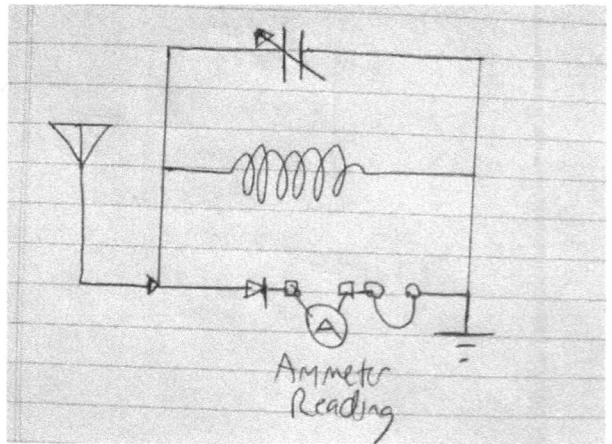

Ammeter Reading

Procedure :

- The procedure follows the construction and testing of the Radio as described in the Set Up.
- With the Radio operational do the following :
- Listening with your earphone find 2 or 3 radio stations and write down their call letters and frequency in the data table.
- For each radio station do these steps to measure the amount of voltage and current in your system as caused by the radio station signal through your radio :
- For both the Voltage and Current readings you can use a digital or a needle-based multimeter. In either case, it is set in DC settings.
- Note that the overall readings are strongly connected to the distance of the radio station signal source and its strength. The farther the source, the smaller the readings.
- Note that for Voltage and Current there are different set up configurations needed to measure the power of the radio signal.
- In both the Voltage and Current cases, the values are very small.
- The DC Voltage setting should start at regular voltage and then dial it down with each step to the millivolt setting.
- For Voltage it is set up across the earphone wires so that it is in parallel with the system.
- Typical values are from 15 mV to 225 mV (but depends on distance).
- Disconnect the multimeter and change the settings to DC Current reading.
- You can start with Amps, but it will quickly go to milliamps (mA) and then microamps (μA) readings.
- Next disconnect the ground wire connection of the earphone and connect it in series to the circuit the multimeter and then to the ground so that the multimeter is now part of the continuous circuit.
- As noted above, start with Amp readings, then move to milliamp readings and then to microamps readings.
- The longer the antenna will also affect the reading values, but they can be from 2 μA and up to 200 μA.
- Record both the Voltage and Current readings for each of the Radio stations discovered on your radio.
- For calculations (being wary of the proper decimal placement) calculate the Power of the Radio Signal in each case.
- For further investigation, find out how far away the stations are and the transmission power of the signal to see how this compares to your calculated power values (that is to say are similar signal transmission emissions at different distances different in power – the farther the smaller the power value, for example?)
- Other things to try : Change the number of coils in your Coil or the size of your Capacitor. Change the length of the Antenna. Try other materials as well.

Data :

Length of Antenna Wire (L) : _____ (m)
Number of Coils on Inductor (N) : _____

Trial	Radio Station (λ or station)	Voltage (mV)	Current (μI)
1			
2			
3			

Calculations :

Be sure to use your Slide Rule!
Measured Power for a Station received by the Radio :

P = V*I

Conversions Needed :

μ means 10^{-6}
m means 10^{-3}
1 m = 100 cm

$1 V = 1 \times 10^3$ mV
$1 A = 1 \times 10^6 \mu A$

Conclusion :

What values did you measure and compute for the power of stations received by your radio? Were they significantly different from each other? If so, why do you think this is the case? (Look up the radio stations and find out how much power they use in broadcasting their signal and how far away from you they are). (Note : Most are 50 kW).

With regards to the radio – you can try different size variable capacitor systems, a different number of coils on your inductor and/or a different length of an antenna. Do any of these affect the performance of the radio and the amount of power measured from your readings for a given station?

Notes : Power readings will be very small and can be from 1×10^{-5} to 1×10^{-8} W (I have personally done this with the items noted above in the photos and measured 2.7 mA and 67 mV for a power reading of 1.8×10^{-7} W)

Activity #25
Make and Measure a Magnetic Field with your Magnetometer Activity
Grade Level : High School
Math Level : Challenging

Make and Measure with a simple Gauss Meter Activity

For information on Magnetic Fields look to other Activities, such as Electric Charge Activity #3, Electromagnet Activity #11, and Electric Motor Activity #13 to read and find the connection of Electricity and Magnetism.

In the case of this Activity, though we are using some basic materials, such as a 9V battery and a Multi-meter, we are using a number of specialized, though inexpensive and easy to find, other electronic parts (many can be found online or at electronics stores).

The essence of what we are doing is building and using what may be called a Gauss Meter or Magnetometer. This is a device to measure the strength of a Magnetic Field along with its direction. It will give us measurements in Voltage (millivolts in this case) which we convert with a convenient formula to Gauss (a common unit for Magnetic Field Flux Density strength). We could use another common unit for magnetic field strength, the Tesla, but it is very large. There are 10,000 Gauss in one Tesla. We can determine magnetic field direction by whether the voltage is positive or negative. If it is found that North is positive then South is negative.

The circuit we construct uses a 9V battery where we control the voltage with a Voltage Regulator so that we have no more than 5V coming into our circuit. Next is what is called the Hall Effect Device which responds to magnetic fields and causes changes in electrical current which effects the voltage readings we make. These Hall Effect Devices are in many electronic devices and even our cars. We will measure changes in voltage as the Hall Effect Device is brought in contact with different magnetic field strengths (recommended to use ceramic magnets).

Purpose : To use a multi-meter with a simple electronic component circuit to measure the relative strength of various tested magnetic fields and then to calculate their strength

Materials :
- 9V Battery,
- Digital Multi-Meter (DC voltage setting used here),
- IC Breadboard,
- Jumper wires to be used with breadboard (4 needed),
- 7805 Voltage Regulator (used to reduce the 9V battery to 5V for use),
- Hall Effect Device,
- 9V battery clip to connect to breadboard,
- Magnets to test (refrigerator, best to use ceramic and other strong ones),
- Slide Rule

Note : This Activity requires patience and some persistence in order to set it up and make it operate, some of the parts are quite small, so Reading Glasses or a Magnifier might be useful. Note too that special electronic parts are needed – many are found online and there may be a nearby electronics store.

Note : Since there are electronics and small parts – it is recommended to have parental permission and supervision in doing this Activity.

Set Up :

1. In order to do this Activity (as with the others) the Set Up is essential. Here we are constructing the circuit to be used to make the device to measure magnetic fields.
2. The essential materials are all listed in the Materials list and are needed. If unsure as to what one is look at the photos when looking them up online or consulting with an electronics salesperson at the store.
3. First a Breadboard is needed. This is the rectangular piece with many holes which is where the components are connected.
4. I have orientated mine so that when looking down at it I have A through J across the top and the numbers start at 1 and go down towards me. The power strips are on either side of the board. This means A1 is in the upper left hand corner.
5. In the following table is the configuration I have first used to set up the circuit when reading or building it from left to right. Also I have started at the top. When looking at the photos it is clear I have moved the arranged pieces to the midsection of the breadboard. I merely redid the experiment. When examined carefully you can do it too since all of the parts are in the same arrangement with respect to each other, only in different holes on the board.

Item	Location
7805 Voltage Regulator	A13, A14, A15 (note it faces into the board)
9V battery lines	B14 (ground), B15 (power)
Jumper Cable 1	D13 connected to G13
Jumper Cable 2	D 14 connected to G14
Jumper Cable 3	H14 to be connected to Multi-meter
Jumper Cable 4	I15 to be connected to Multi-meter
Uncalibrated Hall Effect Device	J13, J14, J15

Photos :

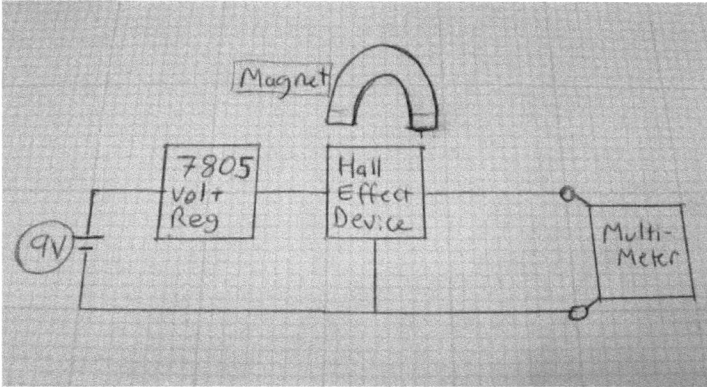

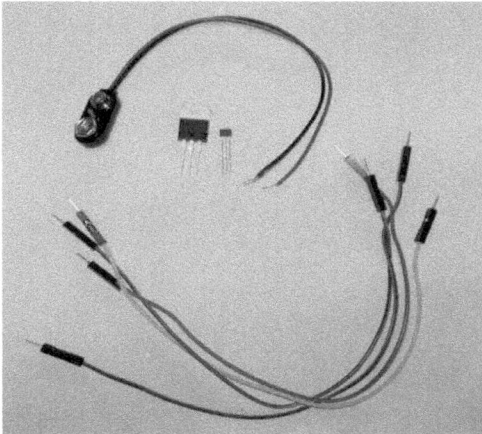

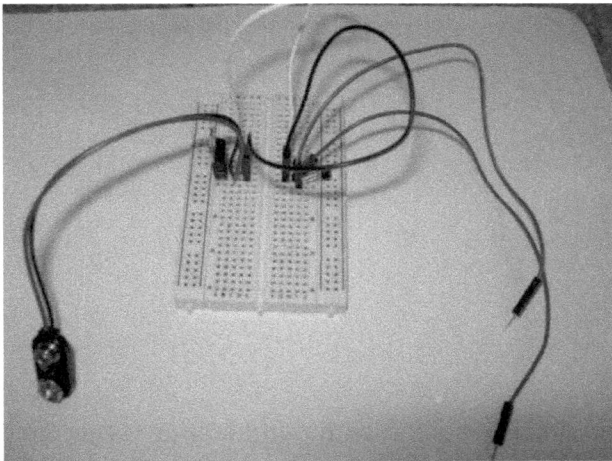

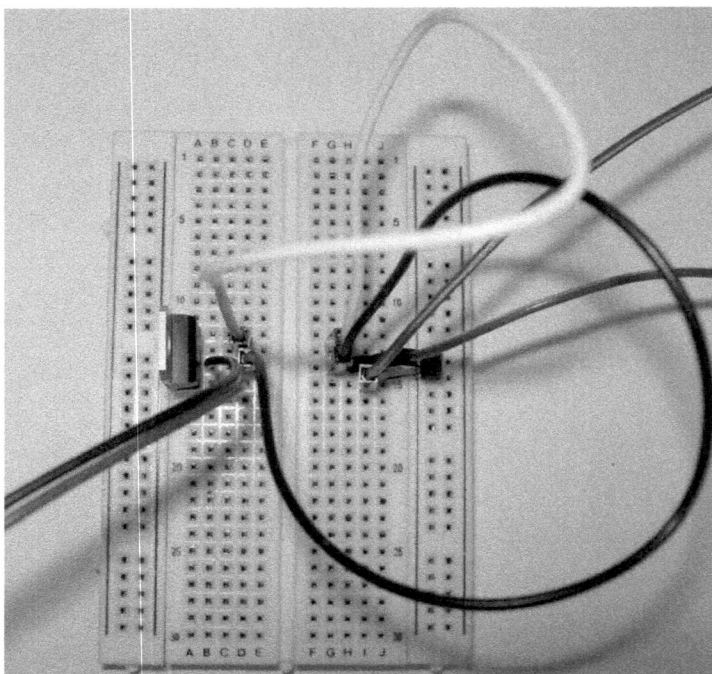

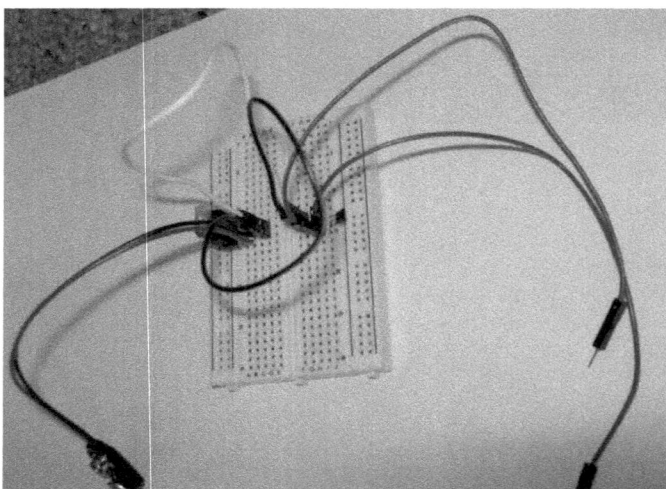

Procedure :

1. Be sure to have all the correct parts for this Activity (along with parental permission and supervision). Read through and complete the Set Up before the Procedure since that is where the magnetic field measuring circuit device is created with your parts.
2. Before testing your device have your multi-meter set and ready to be used. It should be set on DC Voltage and can read 20 V or less (in fact most of your readings will be in the millivolt range (mV)). At this point you can connect the 9V battery to its clip.

3. Since most multi-meters have pin leads to touch with, it is recommended to have clips (see photo) so that they can be attached to the constructed device at the designated loose jumper wires to be touched. It will not matter which touches whichever jumper cable, since it will only ultimately yield a positive or negative answer result which indicates direction of the magnetic field.
4. Now with your assembled, you can test the device (even with no magnets nearby it will read some magnetic field variation, then bring one to the Hall Effect Device and note the changes to the multi-meter readings). You may note that the readings will have a continual fluctuation in the latter digits, but be sure to concentrate on and write down the initial digits (typically the first two) for measurements.
5. With the magnetic field measuring circuit device ready to go, decide on the number of magnets to test, hence the number of trials to conduct. It is best to use a small stack of similar magnets, one at first, then two of them, and so on.
6. Go through your series of Trials. In each case write down the initial reading of the Voltage (V_0) before testing that Trial's magnet(s) to be tested, then the final voltage (V_1) of the reading once the magnet is in place. Be sure to have all of the trial samples at the same distance in each trial.
7. Once all readings are done, disconnect the 9V battery from the system.
8. Calculations :
9. First determine the difference of the Initial and Final Voltages ($V_0 - V_1$).
10. Use the difference in the Magnetic Field Flux Density formula to determine the number of Gauss your magnetic field strength is for your magnet.
11. If you like convert your Gauss measurement to Tesla. Compare your results to each other and to other known values in the Table below in Calculations.
12. Other calculations can be Slope of Magnetic Field Strength vs. Number of Magnets as a possible graph, or perhaps Average Magnetic Field Strength for the measurements you have taken.
13. Oftentimes it is good to redo the Activity to verify results and perhaps try different variations on the same test. Have fun. Enjoy :)

Data :

Trial	V_0 (no magnet present reading) [mV]	V_1 (Magnet present & tested reading) [mV]	ΔV (abs. value of difference of V_0 & V_1) [mV]
1			
2			
3			

- Note that it is not important if the sign of the difference is left in the equation. The sign actually merely indicates the direction of the magnetic field. For example if it is found that North is +, then − results indicate a South pole of the magnetic field is nearest the sensor

Calculations :

Be sure to Use Your Slide Rule! :)

Difference in Voltage Readings DV :

$$\Delta V = [\ V_0 - V_1 \ ]$$

Magnetic Flux Density B :

$$B = \frac{10^3 * (\ \Delta V \ )}{k}$$

k = 2.50 mV/G (unless otherwise noted if using a calibrated Hall Device)

Slope :

$$m = \frac{\Delta Y}{\Delta X}$$

Average :

$$x_{ave} = \frac{\sum x_i}{n}$$

Constants & Conversions & Comparisons :

1 mV = 10^{-3} V = 0.001 V
1 T = 1 Wb/m^2 (T = Tesla, a measure of magnetic field flux density)
1 T = 10^4 G (G = Gauss, cgs units for magnetic field flux density)
1 T = 1 Wb/m^2 = 1 V*s/m^2 = 1 N/(A*m) = 1 J/(A*m^2) = 1 kg/(A*s^2)
1 T = 10^6 μT
(V = Volt, J = Joule, N = Newton, A = Ampere, m = meter, s = second)
Some Magnetic Fields :
A common refrigerator magnet : 5 mT
The surface of a common neodymium magnet : 1.25 T
Earth's average magnetic field strength at surface : 31.8 μT
    (25-65 mT = 0.25-0.65 G)
Sun's average magnetic field strength at photosphere : 50-400 μT

Conclusion :

As you changed both the type and number of magnets, what results did you find? What about the sign (+ or -) of the voltage in your readings – did this change when flipping a given magnet over? Did more magnets increase the strength of the measured magnetic field?

Activity #26
Study of a Flashing or Blinking LED simple Circuit Activity
Grade Level : High School
Math Level : Challenging

In the following Activity, we are using various circuit components to construct, test, measure, and compare the results of the outcome to idealized formula predicting the outcome. In this case our circuit is a Flashing or Blinking LED simple Circuit. Our circuit uses a number of conventional circuit components we can find on the site in various Activities either in use (resistors in Ohm's Law) or discussed in a similar form (such as the Capacitor one) along with a LED, a 9V battery, some wires, and a couple of new items, a Bread Board (a convenient base to construct the circuit on and no need of soldering), and what is called a 555 Timer IC. The 555 Timer IC not only allows the components to be connected as they should be but also it is a transistor that allows a time delay for our circuit. The goal of our circuit is to operate a flashing or blinking LED. With various parts, the timing of the circuit, hence the rate of flashing will change with changes in the resistors and the capacitor values.

Why create a simple flashing LED? The first goal is to compare expected to experimental results in the rate of flashing. The next possible goal is to consider what can a flashing light with known amount of time between flashes could be used for. Consider it perhaps as a timer! Instead of a stopwatch you can use the number of flashes to measure some other Activity that you do – this is your version of a watch basically. Finally it is an exercise in circuit design, development, testing, and use. This could be a new area of interest, even a hobby to explore. Enjoy! :)

Purpose : To construct a circuit from determined electronic parts that results in a flashing LED light where there will be a comparison of calculated predicted time per flash to calculated experimental measurement of time per flash

Materials :
- 9V Battery,
- 9V Battery clips,
- Resistors : Various – see Sets for suggestions,
- Capacitor : Various – see Sets for suggestions,
- 555 Timer IC,
- Jumper Wires for Bread Board (at least 3 needed),
- LED :
- Bread Board,
- Graph Paper,
- Stop Watch,
- Slide Rule

Note : As with most Activities, it is best to have parental permission and supervision in this one. This Activity requires a number of specialized electronic components (resistors, capacitors, et al) so it will require some shopping at electronics stores either online or in your area if available.

Note : The person undertaking this Activity needs to have patience and persistence as it may take a number of tries (it did for me).

Note : In the Set Up I have a recommended set up, but realize one could place the circuit in different locals on the Bread Board, as long as it is in the same analogous arrangement.

Some Possible Sets of Resistors and Capacitors to be used in the Circuit :

Note : The following suggestions are possible sets to use for your blinking LED circuit, but it is suggested to try other values as this will affect the rate of blinking. – Realize that each set is its own unique set of data to be analyzed.

Set	Resistor 1 (R1)	Resistor 2 (R2)	Resistor 3 (R3)	Capacitor
Set 1	33K	100K	1K	10μF
Set 2	1K	470K	1K	1μF
Set 3	1K	100K	1K	100μF
Set 4	10K	1K	1K	10μF

Set Up :

- Note : Below is a step-by-step Table for each of the components of the flashing LED circuit. Recognize that you can place these components elsewhere on the Bread Board, but be sure to change all of the corresponding points for your new configuration.
- At the start I have the Bread Board in front of me with A1 in the upper left corner (as seen in the photographs). You can see A through J across the top and 1, 2, et al down the sides.

Step	Item	Position
1	555 Timer IC	The 555 Timer IC has its notch at top away from you. One side has its connections going from E2 through E5 while the other side has its connections going from F2 to F5. For certainty, post 1 is in E2, post 2 is in E2, while post 5 is in F5 and F4 is post 6
2	Jumper Wire 1	D3 is connected to G4 (post 2 to post 6)
3	R1 (Resistor 1)	C3 is connected to G3 (post 2 to post 7)
4	R2 (Resistor 2)	I3 is connected to Power (red) (post 7 to Power)
5	Capacitor	C3 is connected to Ground (blue/black) (post 1 to Ground)
6	Jumper Wire 2	C2 is connected to Ground (post 1 to Ground)
7	Jumper Wire 3	I2 to Power (post 8 to Power)
8	LED	D3 is connected to C15 (any open spot) (post 3 to Open Spot)
9	R3 (Resistor 3)	B15 to Ground
10	9V wires on 9V clip	Black wire is connected to Ground (blue/black) Red wire is connected to Power (red)

- With the completion of all of these steps it is best to test your Flashing LED circuit. Check and redo it as needed if not working.

Photos :

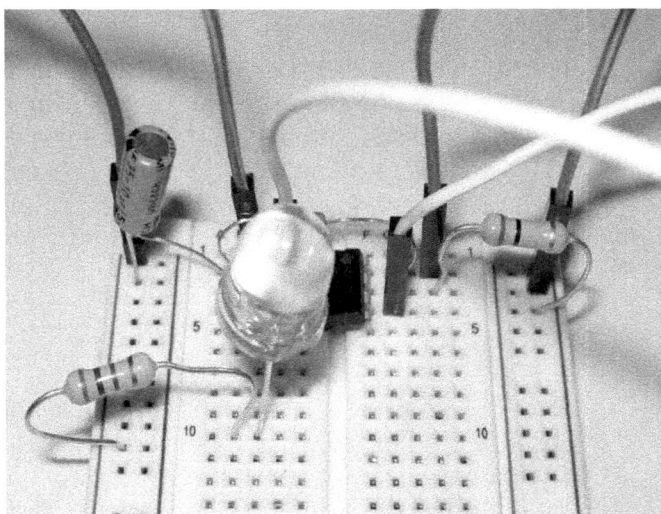
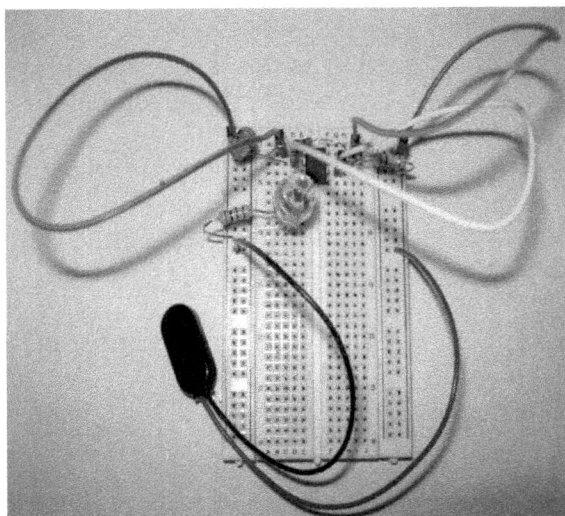
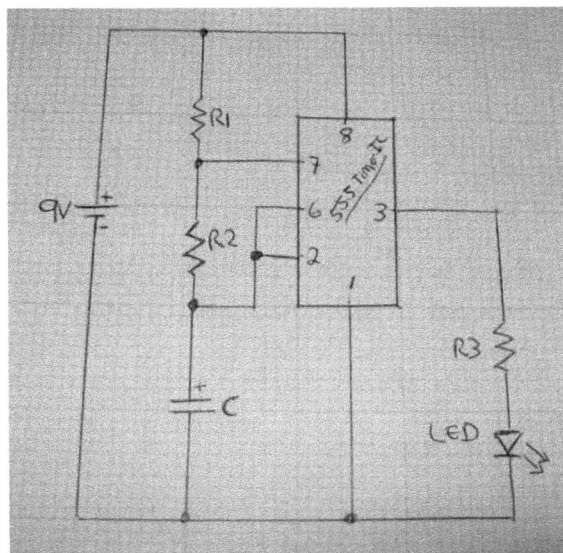

Procedure :

1. At this point you have all of the needed materials, have read through the directions, and completed the functioning circuit from the directions in the Set Up.

2. For each experiment you are using a given Set of materials. Choose how many of the Methods noted in the Data section you want to do. It is suggested to do all 3 in succession as they are easy to do. Directions for each and the Calculations for each section are then noted within the body of the directions for the given Method.

3. There is an important thing to consider before going on. In each of the Methods, you will have to keep your attention to both Time and the Number of Flashes of the LED circuit. Clearly mistakes by one person doing the readings can be made easily. To avoid problems, find solutions to this problem, such as having two people do the Activity, where one is the Timer and the other is the Flash counter. Another one can be having a Timer that has an alarm or countdown with signal system to let you know time is up.

4.

5. Determining the Predicted Time for each flash (P) :

6. Before we address the data from our circuit and our Studies it is best to find what we expect from our circuit in terms of the time for each flash (P) – this is what we are going to compare to.

7. To calculate the time per flash (P) we need to know all the values of the components of our circuit and plug them into the given formula in the Calculations section.

8. The two times we must determine are the amount of time of the capacitor charging (T_C) and the amount of time of the capacitor discharging (T_D). The total time per flash (P) is merely the sum of these values. Calculate this – be wary of the exponents involved).

9. This is done for each Set you use.

10.

11. Method 1 :

12.

13. In this first study of the flashing LED we will have a set amount of time we are going to observe the flashes. I have written down 10 seconds, but you can choose another value if you wish.

14. In order to take a reading, activate the circuit so that the LED is flashing. Have your stopwatch handy.

15. It is best to have two people do this task, but perhaps you have a countdown timer that signals you when the time is up.

16. At a given flash, now activate the timer, do not count this one as this is the starting signal in essence. Every flash after this for the next 10s (or whatever time you have chosen) is counted.

17. At the end of the time cycle, write down your Number of Flash (N) count in the data table.

18. Do this for at least 3 Trials (you can do more if you wish – it is often a good practice to have a practice run that is not counted to see how it goes).

19. Calculations for Method 1 :

20. Determine the Average (Nave) of the Number of Flashes (N) from your number of Trials (n).

21. Use this determined Average value to find the Time per Flash (P).

22.
23. Method 2 :
24.
25. In the second Method we will do at least 3 Trials, but it is often a good idea to do more (5 is a good suggestion).
26. In this Method we will set an established amount of time (5 seconds, then 10 seconds, and so on) for each Trial – during that time frame we count the number of flashes (N) that occur and record these results.
27. This is definitely a Method where two people are a very good suggestion at doing the Method. One person does the timing and calls out when time is up where the flash counting person then stops and records those results.
28. Notice the Data Table – it has only one trial for 5 seconds, one trial for 10 seconds and so on, but this is not the best way to do this Method. Much like Method 1, you should do each trial 3 times (3 times for 5 seconds where the flashes are counted). Record all the counted values for a given time and take the Average for this number of flashes as you did in Method 1.
29. Calculations for Method 2 :
30. Either you have the number of flashes for 5 seconds or its average – whichever it is – this becomes one of the coordinates on a x-y graph.
31. To Graph the data, we will do something rather unconventional but useful in our analysis. Here we will place the Independent variable, the no. of seconds (5, 10, 15, et al) on the y-axis and place the Dependent variable the corresponding no. of flashes (N or Nave) on the x-axis. When done you have minimally 3 points (5 is best).
32. Draw a best fit line through these points and use the origin (0,0) as the intercept point.
33. Determine the Slope of this line. The slope is the amount of time per flash (P).
34.
35. Method 3 :
36.
37. In the third Method we will do at least 3 Trials, but it is often a good idea to do more (5 is a good suggestion).
38. In this Method we will set an established number of flashes (5 flashes, then 10 flashes, and so on) for each Trial – during those flashes, with the first call the first '0' at the start of the timer and then count the number of flashes for that trial and stop the timer with the last flash for that trial and record these results.
39. This is definitely a Method where one person can do the work, but if desired use two people if you wish. One person does the counting and calls out when both starting and when done counting so that the timer person when to start and stop the watch.
40. Notice the Data Table – it has only one trial for 5 flashes, one trial for 10 flashes and so on, but this is not the best way to do this Method. Much like Method 1, you should do each trial 3 times (3 times for 5 flashes where the time for each trial is measured and recorded). Record all the values for a given time and take the Average for this time as you did in Method 1.
41. Calculations for Method 2 :
42. Either you have the amount of time for a given number of flashes or the average time – whichever it is – this becomes one of the coordinates on a x-y graph.

43. To Graph the data, we will do the conventional of graphing the Independent variable on the x-axis which is the number of flashes (N) while the time or average time for the given number of flashes as the Dependent variable y-axis coordinate of this pair. When done you have minimally 3 points (5 is best).
44. Draw a best fit line through these points and use the origin (0,0) as the intercept point.
45. Determine the Slope of this line. The slope is the amount of time per flash (P).
46.
47. Final Considerations with your Data and Calculations :
48.
49. For all the Methods used (best to do all three) – examine your calculated time from your experiments for the amount of time per flash (P). How does it compare to the expected time as determined from the predicted values in the formulae used at the start of your calculations? To determine this, use the percent error formula, it will give you a good idea as to how close or similar your values are to the expected values.

Data :

Method 1 :

Time (t) = 10s

Trial	No. of Flashes (N)
1	
2	
Ave.	

Method 2 :

Trial	Time - No. of Seconds (s)	No. of Flashes (N)
1	5	
2	10	
3	15	

Method 3 :

Trial	No. of Flashes (N)	Time for no. of flashes (s)
1	5	
2	10	
3	15	

Calculations :

Be sure to Use Your Slide Rule! :)

Formulae for Data (To determine Experimental Time of Flash) :
Average :

$$\mathbf{X_{ave}} = \frac{\sum x_i}{n}$$

Slope :

$$\mathbf{m} = \frac{\mathbf{\Delta Y}}{\mathbf{\Delta X}}$$

Time for Flash (Period P) :

$$\mathbf{P} = \frac{t}{N_{ave}}$$

Formulae for Predicted Time of Flash Outcomes :

Time for Capacitor Charging :

$$\mathbf{T_C = 0.69 * (R_1 + R_2) * C_1}$$

Time for Capacitor Discharging :

$$\mathbf{T_D = 0.69 * R_2 * C_1}$$

Time for Complete Cycle :

$$\mathbf{P = T_C + T_D}$$

Formula to compare Experimental to Predicted Outcomes :

$$\mathbf{\%E} = \frac{\mathbf{[\ Predicted\text{-}Experimental\ ]}}{\mathbf{Predicted}} * \mathbf{100\%}$$

Conclusion :

How did your calculations for the experimental time per flash compare to the predicted complete cycle time for flash you measured? How similar were your experimental times per flash to each other for all three methods? How similar were your slopes for experimental time for flash to each other using the two graphing methods?

Activity # 27
Solar Cell Examinations Activity
Grade Level : High School
Math Level : Calculating

Solar Cell Measures Activity :

This Activity has the following Prelude on Solar Energy and Photovoltaic Cells for those interested in the subject. Otherwise you can move ahead to the Activity that follows concerning an investigation of these photovoltaic cells.

The Sun provides an enormous amount of energy for the Earth. The energy is discussed with more details and numbers in the Solar Constant Activity #27, but it can be said that the Sun's energy reaching the Earth's upper atmosphere is on the order of 174 petawatts of power (peta meaning 10^{15}). This energy drives the wind, the waves, provides heat, light, and energy for life forms of the planet.

Solar Energy in many forms, especially in the present world of technology, is used for many reasons. The most obvious is light, which is utilized for everyday visual uses. The most obvious employment of the Sun outside of regular light and heat is the growing of crops by humans which takes advantage of photosynthesis by plants which converts carbon dioxide (CO_2) and water (H_2O) into molecules to provide energy, namely sugar. The heat of the Sun has a myriad number of effects and applications in our everyday lives and is the primary driver of weather (both the wind and water) the globe over and is needed by humans.

Humans have basically utilized light in two ways categorically : passive and active solar energy technologies. These classifications result from the way the sun's energy is captured, converted, and distributed. The Passive Solar Energy Technology will be explored in more detail in the Rate of Cooling Activity #28 indirectly, but it can be said here that it is based upon the thermal properties of materials positioned in favorable ways to take advantage of the Sun' energy, heat, and light.

Active Solar Energy ideas are ones that take the sunlight and convert it into other forms of energy (the most common being electricity) for use. In this technology there are two chief types : 1) concentrating solar power - focused sunlight and the use of Stirling engines or for boiling water to use with steam turbines 2) utilizing the photoelectric effect - photovoltaics cell use (which is the focus of this Activity).

In concentrating solar power, here optics such as lenses or parabolic mirrors are used to focus the sun's light energy onto a region or point to act as a heat source to actively power a power station, such as a Stirling engine. These heating systems often heat a fluid of some form to cause it to do work, which can then activate an electric generator. Others can be used to boil water to turn turbines as well.

The other type of Active Solar Technology are Solar Cells –aka Photovoltaic cells (the latter term can apply to any and all light whereas the former applies to sunlight) which convert sunlight directly into electricity through the photoelectric effect.

The term 'photovoltaic' comes from 'photo' Greek for 'light' and 'voltaic' derived from the Italian physicist Volta and the term 'volt' came from which is the unit of electro-motive force, so 'voltaic' is intended to mean 'electric'. This idea was first written about but not constructed in 1839 by the French physicist A. E. Becquerel. The first created one came in 1883 by Charles Fritts, who used coated the semiconductor selenium with a thin layer of gold at the junctions. This first one was on 1% efficient. Not long after the first solar cell based on Heinrich Hertz 1887 photoelectric effect came from the Russian physicist Aleksandr Stoletov. Even Albert Einstein played a part in the history of the solar cell with his 1905 paper explaining the photoelectric effect for which he won a Nobel Peace Prize in 1921. In time more research and modifications were made to the solar cell. A much more efficient form using a diffused silicon p-n junction was developed by Chapin, Fuller, and Pearson in 1954. Some solar cells today have reached values of 24%, and even 42% absorption efficiency (the latter one with concentrated light), well above the average in the industry which is at 12-18%. (Note that this stands in contrast to sunlight-to-electricity efficient, which is lower). Today there are even Megawatt solar power generating plants being built.

This type of electrical energy production is not only found in small devices, such as arrays for charging batteries and uses in calculators, but is always in use in space (satellite and space station power sources) and it is finding a growing market worldwide here on the ground. The efficiency of them is increasing, their fragility is decreasing, and their costs are coming down. There are even bendable solar cell arrays that are used as shingles on roofs these days. As of

2010 power generated by them is going on in over 100 countries. It is estimated that 4800 GW could be harnessed this way yet only some 21 GW are being used at present. Since 2002, photovoltaic production has risen by 20% per year making it the fastest-growing energy technology. Historically the first commercial use of it was in 1966 on Ogami Island in Japan to change the Ogami Lighthouse from gas torch to fully-sufficent electrical power. The three leading countries in use of this technology are Japan, Germany, and the United States. Presently most stations range from 10-60 MW capacities and in the near future 150 MW or more are projected. All of these worldwide efforts are driven by the need for renewable energy sources and their utilization.

Solar Cells are made of semiconducting materials (these are elements that are in the metalloid region of the Periodic Table between metals and nonmetals). Over the cell is a clear covering of glass or plastic to allow light in, but to protect the semiconducting wafers. These cells are encapsulated and care connected in series and/or parallel connections depending on the need for greater increases in voltage and/or current and for the overall power of the system. The arrangement is called an array.

The solar cell generates a DC (direct current). The power output is measured in watts or kilowatts (depending on the number of cells in the array). To determine the number of cells needed and the arrangement of cells needed, the energy needs calculated in watt-hours, kilowatt-hours, or even a energy per day as kilowatt-hours per day are often used. A quick mental rule is this : Average Power is equal to 20% of Peak Power.

The power from a solar cell array can be fed directly into the electric power grid through inverters. If these arrays are stand-alone the energy is often stored in rechargeable batteries if it is not immediately being used (such as powering a device). Smaller panels can be used as chargers and direct power sources for cellular phone chargers, solar-powered calculators, solar lights for things like bikes and solar-charged camping lanterns.

How do they work? When photons encounter the surface, some pass right through (since their energy has no effect on the materials), some reflect off the surface of the cell, while some photons have the right amount of energy that corresponds to the 'silicon band gap'.

These photons that hit the solar panel and are absorbed by the semiconducting materials energize the electrons in the valence band. These electrons are regularly tightly bound in covalent bonds between neighboring atoms. The energy they are given excites them into the conduction band where it can move more freely in the semiconductor. Where the electron was, now has what is called a 'hole'. This creates a potential for electrons in surrounding molecules to move into it. Hence electrons move one way while the holes move in the more-or-less opposite direction. Next these negatively-charged electrons are knocked loose from their atoms are allowed them to flow though the semiconducting material and produce electricity. The cells composition only allows for the electrons to move in one direction. Hence, when cells are connected to each other will have an overall voltage and current flow and produce a measurable and useable amount of DC electricity.

The most commonly known solar cell configuration is called the p-n junction and is made from silicon laced with other semiconducting materials. Basically it is a materials where a layer of n-type silicon is placed in direct contact with a layer of p-type silicon. That is just a mental model, while typically the layer of silicon has one side diffused with a n-type dopant into a p-type wafer (or vice versa) so that the effect is the same.

In the p-n junction type there is a diffusion of electrons from the region of high electron concentration (the n-type side of the junction) into the region of low electron concentration (the p-type side of the junction). When the electrons diffuse across the p-n junction, they recombine with holes on the p-type side. This does not continue indefinitely, since a charge build up results in an electric field, which in turn, creates a diode that promotes charge flow (aka drift current) which opposed and eventually balances out the diffusion of electrons and holes. The area that no longer contains any mobile charge carriers is called the space charge region.

The solar cell is connected to an external load at its positive and negative points. The voltage measured is equal to the difference in the electrons in the p-type portion and the holes in the n-type portion. With voltage, when connected to any load (which has resistance) then a current can occur (Ohm's Law Activity #10).

The amount of Power from a solar cell easily can be seen from the amount of light falling on it, which is one of our Activities. Also, the number of solar cells used and how they are arranged will affect the amount of power available from a solar array. The two common types of arrangements are called : Series & Parallel arrangements.

Like in the Ohm's Law where Resistors are connected one after another and hence are a Series, the same is true for solar cells. Here the positive lead connects to the next solar cell's negative lead and so on. What is left is the first one has a negative lead not connected, and the final one has a positive lead not connected. These can be connected to a device or a multimeter for determining its readings. One should find that the voltages should sum up and hence increase, while the current values should remain constant.

In the Parallel arrangement, the positive leads are all connected as are the negatives to each other in two lines across all of the solar cells. From the lead or final solar cell to additional lines are connected for connection to a device or a multimeter. Here all the voltages are the same, so it will not increase, while the currents will all sum up and have a larger value.

One could also be more creative, possibly wanting both greater voltage and current and create, for example, two separate lines where each line has two solar cells in series with each other while the two lines are in parallel to each other. This system will add up in a way so that it has both increases in voltage and in current values.

Our investigations will take the solar cell and concentrate on constant light sources at different distances, angles, and even the amount and/or type of light reaching the solar cell. The ast activities examines the connections of more than one solar cell and looks at series and parallel arrangements.

One idea not explored here is distance of the light source from the solar cell, since this addressed in the Inverse-square Law of Light Activity. Also the solar cell can be used in the Solar Constant Activity #27.

The Solar Cell Investigations :

I. **Solar Cell Power due to Exposed Area of Cell**
II. **Solar Cell Power due to Angle of Light Exposure**
III. **Solar Cell Power due to Light Intensity**
IV. **Solar Cell Power due to Wavelength of Light**
V. **Solar Cell Power when connected in Series**
VI. **Solar Cell Power when connected in Parallel**

Activity I : Solar Cell Power due to Exposed Area of Cell
Purpose : To measure and determine the effect on voltage and current measurements (to calculate power) for a constant light source at a given distance while changing the amount of solar cell surface area exposed to the light.

Activity II : Solar Cell Power due to Angle of Light Exposure
Purpose : To measure and determine the effect on voltage and current measurements (to calculate power) for a constant light source at a given distance while changing the angle that the light strikes the solar cell.

Activity III : <u>Solar Cell Power due to Light Intensity</u>
Purpose : To measure and determine the effect on voltage and current measurement (to calculate power) for a constant distance light source of varying intensity light emitted and received by the solar cell.

Activity IV : <u>Solar Cell Power due to Wavelength of Light</u>
Purpose : To measure and determine the effect on voltage and current measurements (to calculate power) for a light source emission filtered by color filters to allow certain wavelength access to the solar cell.

Activity V : <u>Solar Cell Power when connected in Series</u>
Purpose : To measure and determine the effect on voltage and current measurements (to calculate power) when a solar cell array has its cells connected in a series circuit manner when receiving a varying light source intensity.

Activity VI : <u>Solar Cell Power when connected in Parallel</u>
Purpose : To measure and determine the effect on voltage and current measurements (to calculate power) when a solar cell array has its cells connected in a parallel circuit manner when receiving a varying light source intensity.

Materials :

- Three 1.5 V Solar Cells (only one if not doing series or parallel),
- 6 Alligator Clip wires or wires to connect the solar cells (2 for one),
- Multimeter,
- Bulbs : 25 W, 40W, 60W, 75W, 100W,
- 2 Small Lamps (regular post type and flexible head),
- Protractor,
- Meter Stick or Measuring Tape,
- Color Filters (R, G, B),
- Piece of Dark construction paper or cardboard,
- Scissors,
- Tape,
- Graph Paper,
- Small penlight to read and write by,
- Item(s) to create angle for solar cell to rest on and be at – can be a stack of books/magazines, use of poster board or cardboard folded, or some other imaginative item to get the job done,
- Slide Rule

Procedure :

1) All Activities use a Light Source (Lamp with at least one bulb used), the multimeter, and at least one solar cell, and all have their calculations done with the slide rule.
2) Recommended : Use rectangular, 1.5 V solar cell
3) Note that since the effect of the light on the solar cell is the primary measure of interest, try to do most of the measurements with as little external light as possible. (have the small light available to write things down).
4) Exercise safety in using the lamp : Do not touch the bulb, Leave it unplugged until the bulbs are properly in place and then activate it, do not place items near or on the bulb (such as the solar cell or multimeter).
5) It is best between trials to turn off the light, turn off the multimeter, and then set it up to operate for that given trial.
6) Recognize that not only do you turn the dial (typically) to change from reading voltage to reading current on the multimeter, you must also change the configuration of the test wires. Be sure of the instructions on how to properly operate the multimeter for voltage and current readings.
7) Since the solar cell is part of every experiment, always have the positive and negative leads coming from one (or more) the solar cell(s). If properly done, there are only two wires, one positive and one negative to touch or connect to with the multimeter leads.
8) Note : Lamps are used so as to control the amount of light, its distance, angle, et al. But once there is an understanding of the results in using a solar cell, one coud use their intended source, namely the Sun on sunny days and try these experiments using it when and where possible. Be sure to not look directly into the Sun. Use shadows of pencils act as guides for pointing items directly at the Sun. Also employ safety when using bulbs as well – do not touch active bulbs nor even when turned off – do not look directly into the light. Employ common sense safety procedures.
9)
10)   Activity I : Solar Cell Power due to Exposed Area of Cell
11) Measure the Length and Width of the Solar Cell. Calculate its Area.
12) Use the dark construction paper. Measure and draw a rectangle 3 times the width by one length of the solar cell dimensions.
13) Cut the first piece to be the length and width of the solar cell.
14) For the second, cut a rectangle the width and for the length measure 75% of the length and cut it there.
15) Save the 25% piece too since this is the final piece.
16) For the third, cut a rectangle the width and for the length measure 50% of the length and cut it there.
17) Each of the rectangles are used to cover the exposed area of the solar cell in turn as noted in the data table.
18) For this Activity, it is best to use the flexible light since it can be directed onto the solar cell lying flat on the table or floor. If not, just be sure not to move the lamp once in place so that the amount of light available is the same for each trial.
19) In setting up the solar cell, be sure to connect the wires so that you can connect a meter to take readings.

20) Choose a Wattage bulb (60 W recommended) and a distance for the light (about 25 cm recommended). Plug in and activate it.

21) Choose an order (like the one noted in the data table) for percent exposed.

22) Start with the Voltage readings [V]. (note readings might not be in volts (V), depending on distance and bulb intensity)

23) After each reading, record it, turn off the light and place the appropriate cover on the solar cell for the next reading.

24) Reactivate the light and continue the readings.

25) Next configure the multimeter to read Current [I]. Check to see it has the proper setting (note that readings may not be in amps (A) so be sure to correct for this).

26) Note : 1000 mX = 1 X (X can be volts or amps)

27) Go through the process and record all of the current values.

28) Calculate the Power using your slide rule.

29) Create 2 graphs of this : First a Bar graph of Power versus Percent Exposed. Second a Line graph of the same variables and compute the slope of the best fit line for this case.

30) A good follow up to this activity is to try other wattage bulbs to see if the results vary or not.

31)

32)   Activity II : Solar Cell Power due to Angle of Light Exposure

33) In this Activity, use the lamp (60 W bulb recommended), the solar cell, the multimeter, a stack of books and the protractor.

34) The best set up is using a standard post lamp set at a constant distance (25 cm to 50 cm).

35) The solar cell is atop a duel set of books and goes from lying horizontally to standing vertically, propped up by paper/magazines/cardboard.

36) The key is to have the center of the bulb and the solar cell when standing up directly in line (this is referred to as vertical or 90°)

37) The angle that the solar cell is at is measured (see data table below)

38) The angle is with respect to the ground and is the compliment of the angle with respect to the solar cell. Determine this.

39) At each angle, measure the Voltage and the Current using the multimeter.

40) When done, calculate the Power of the solar cell for each of the angles.

41) Create both a Bar Graph and a Line Graph of Power versus the Angle and compute the slope of the best fit line for the line graph.

42) Like the other Activities, try other wattage bulbs for comparison.

43)

44)   Activity III :

45) For this Activity, use the standard post-style lamp, have all the bulbs available, and use the meter stick to determine distance between the lamp and the solar cell.

46) As in Activity 2, you could have the solar cell stand vertically and facing the bulb (though no matter its orientation to the light if held constant it will be okay).

47) Choose a distance between the lamp and the solar cell that is held constant (recommended 25 cm to 50 cm).

48) Set up the lamp with the lowest wattage bulb first and work your way through to the highest wattage bulb for each of the trials.

49) With a given bulb in and on measure both the voltage and current produced in the solar cell and record your results in the table.
50) When done with the data table, calculate the Power for each of the bulbs.
51) Graph in a Bar Graph and a Line Graph the results of the measurements with Power as the y-axis and Bulb Wattage as the x-axis.
52) For the line graph, draw a best fit line and compute slope.
53)

54)   Activity IV : Solar Cell Power due to Wavelength of Light
55) In this activity use one bulb type (60 W recommended) and placed in the lamp at a constant distance (25 cm to 50 cm recommended).
56) Begin with the measurements of the solar cell for voltage and current with no filter in place.
57) Between each trial turn off the light and set up for the next set of measurements.
58) For each of the trials, place one of the color filters (red, green, blue) on the photovoltaic cell and then activate the light and take measurements of voltage and current with that filter in place.
59) For each of the trials, white light and the filter, calculate the Power being produced by that light as measured.
60) Compare results to each other (one way is to let the white light be the 100% base and divide this value into each of the others to find its percentage).
61)

62)   Activity V : Solar Cell Power when connected in Series
63) In this Activity, 3 Solar Cells are connected to each other in a Series Circuit fashion (that is there is one path for the current to flow). To do this : use the alligator clip wires and connect the positive terminal of one solar cell to the negative of the next in line, then its positive to the negative of the next.
64) In the final step have one alligator clip attached to the first solar cell's negative terminal while another alligator clip is attached to the last solar cell's positive terminal. These are the terminals to measure from with the multimeter.
65) Choose a bulb wattage to begin with at a constant distance (25 cm to 50 cm recommended).
66) Choose a way to configure the solar cells so that they receive approximately the same amount of light (examples : 1) they can lie flat in a circle around the light if the wires are long enough, 2) the best choice is to stand them up as we did in Activity 3 since we are going to not only compute this Power but also compare this outcome to Activity 7 when connected in Parallel).
67) Once set up, activate the light and take measurements of voltage and current using the multimeter. For the measurements record the overall circuit and each of the solar cells in turn as well.
68) Note : Be sure to connect the multimeter correctly when measuring voltage and current. For example – the voltage can be done by touching across each set of terminals in the array, while the current is found by connecting the ammeter setting of the multimeter in series with the circuit at one spot (all of the currents will be the same – try a couple different spots to be sure).
69) Also for comparison purposes, after completing the first data set, you redo this exercise with a different size bulb.

70) Before calculation, what does the data show you? (If properly done, when in Series, the voltage should be the sum of all the voltages).

71) From the data set, calculate the Power of the solar cells connected in series. This will be compared to the Parallel circuit case next.

72)

73)   Activity VI : Solar Cell Power when connected in Parallel

74) In this Activity, 3 Solar Cells are connected to each other in a Parallel Circuit fashion (that is there is a unique path for each of the solar cells current to flow). To do this : use the alligator clip wires and connect the positive terminal of one solar cell to the positive of the next in line, then connect an alligator clip wire between the negative terminal of the first solar cell to the second. Connect the third to the second in exactly the same fashion. Hence all of the positives have a line and all of the negatives have a line, much like a ladder.

75) In the final step have one alligator clip attached to the first solar cell's negative terminal while another alligator clip is attached to the last solar cell's positive terminal. These are the terminals to measure from with the multimeter.

76) Choose a bulb wattage to begin with at a constant distance (25 cm to 50 cm recommended).

77) Choose a way to configure the solar cells so that they receive approximately the same amount of light (examples : 1) they can lie flat in a circle around the light if the wires are long enough, 2) the best choice is to stand them up as we did in Activity 3 since we are going to not only compute this Power but also compare this outcome to Activity 6 when connected in Series).

78) Once set up, activate the light and take measurements of voltage and current using the multimeter. For the measurements record the overall circuit and each of the solar cells in turn as well.

79) Note : Be sure to connect the multimeter correctly when measuring voltage and current. For example – the voltage can be done by touching across each set of terminals of any member in the array, while the current is found by connecting the ammeter setting of the multimeter in series with the circuit at one spot – that is be a part of the path for a given solar cell (all of the voltages will be the same – and the currents should be the same too since they are the same solar cell, but this is not always the case - try a couple different spots to be sure).

80) Also for comparison purposes, after completing the first data set, you redo this exercise with a different size bulb.

81) Before calculation, what does the data show you? (If properly done, when in Parallel, the circuits overall current should be the sum of all the currents of the separate solar cells).

82) From the data set, calculate the Power of the solar cells connected in series. This will be compared to the Series circuit case in the last Activity.

83) Notes for all : When constructing Bar Graphs, note that the area of the bar when the axes are Voltage and Current will become Power!

Data :

Activity I : Solar Cell Power due to Exposed Area of Cell

Bulb Wattage Used : _____ W
      Recommended : 60 W

Area of Photovoltaic Cell : _____ cm^2

Distance of Light : _____ cm
      Recommended : 25 cm to 50 cm

% Exposed	Area Exposed (cm^2)	Voltage (V)	Current (A)
100			
75			
50			
25			
0			

Activity II : Solar Cell Power due to Angle of Light Exposure

Bulb Wattage Used : _____ W
      Recommended : 60 W

Distance of solar cell nearest base to lamp : _____ cm
      Recommended : 25 cm to 50 cm

Angle (°)	Angle to Light (°)	Voltage (V)	Current (A)
0 (horiz.)			
15			
30			
45			
60			
90 (vertical)			

Activity III : Solar Cell Power due to Light Intensity

Distance of Light Source : _____ cm
   Recommended : 25 cm to 50 cm

Wattage Used (W)	Voltage (V)	Current (A)
25		
40		
60		
75		
100		

Activity IV : Solar Cell Power due to Wavelength of Light

Bulb Wattage Used : _____ W
   Recommended : 60 W

Distance of Light Source : _____ cm
   Recommended : 25 cm to 50 cm

Wavelength color	Voltage (V)	Current (A)
White		
Red		
Green		
Blue		

Activity V : Solar Cell Power when connected in Series

Distance of Light Source : _____ cm
   Recommended : 25 cm to 50 cm

Wattage Used (W)	Voltage (V)	Current (A)
25		
40		
60		
75		
100		

Bulb Wattage Used : _____ W

Solar Cell	Voltage (V)	Current (A)
1		
2		
3		

Activity VI : Solar Cell Power when connected in Parallel

Distance of Light Source : _____ cm
  Recommended : 25 cm to 50 cm

Note : This table is for the overall circuit values and not for the individual solar ce ls in the array. The table below that is for each of the solar cells.

Wattage Used (W)	Voltage (V)	Current (A)
25		
40		
60		
75		
100		

Bulb Wattage Used : _____ W

Solar Cell	Voltage (V)	Current (A)
1		
2		
3		

Calculations :

Se sure to use your Slide Rule !

Area of Solar Cell : **A = L * W**
(rectangular area : Area = Length x Width)

Slope : $\mathbf{m = \dfrac{\Delta y}{\Delta x}}$

90° = Angle(1)+ Angle(2)

Electrical Power : (P is in Watts)

P = V*I

Conclusion :

The conclusion depends on which of the Activity(ies) were undertaken. In all cases, there is a calculation of Power and it is examined as compared to some other changing variable (light intensity, distance, et al). What do the results show and how do they compare to science-based expectations?

Summary and Alternate Ideas :

Besides using a constant light source as we did here, try and use the Sun. Be sure to understand its changing position throughout the day and other factors that may affect outcome as well. This makes for a good comparison to the light bulbs used as well.

Activity #28
Calculating the Solar Constant
Grade Level : High School
Math Level : Challenging

The Solar Constant, Solar Energy and the Slide Rule Activity –

The Sun is the home star of the Earth and the whole of the Solar System. It is 1.4×10^6 km in diameter (approximately 109 x Earth's diameter), hence its volume could hold over 1 million Earths! The mass of the Sun is approximately 2×10^{30} kg (some 330,000 x that of the Earth).

Despite this the Sun is fairly average amongst the family of stars in the cosmos being classified as a G2V type. (Class ranges : O,B,A,F,G,K,M with each having 10 subcategories and the V is a Main Sequence star meaning it is in the prime of its life and generates energy from fusion of hydrogen into helium).

The Sun is composed primarily of Hydrogen (73.5 %) and Helium (25%) with the remainder being many common elements such as Oxygen, Carbon, Iron, Sulfur, and Nitrogen.

The Sun is not only radiant in visible light spanning the spectrum (ROYGBIV) as uncovered and described by Newton as being the components of white light) but also emits radiations at many other wavelengths, such as Radio Waves, X-Rays, Infrared Radiation, and Ultraviolet Rays.

In this Activity we examine the Sun's energy output directly.
Where does the Sun get all of this energy? Like other main sequence stars, the Sun has nuclear fusion taking place in its core. In this process in the 17 million degree core the Sun converts 630 billion kg of Hydrogen into 625.7 billion kg of Helium (a difference of some 4.3 billion kg of mass loss) per second. This results in a power of 3.9×10^{26} W given off by the Sun! In time through the processes of conduction, convection, and radiation, the energy reaches the surface in some 20,000 years and is emitted into space as electromagnetic radiation. The photons (all of which that move at the speed of light 3×10^8 m/s) that reach the Earth (some 1.5×10^8 km) take over 8 minutes of time.

The importance of the Sun's energy cannot be overstated. It is the primary reason that the Earth is at the temperature it is at. The Sun drives our atmosphere and weather systems. Hence it is the energy source for all wind generators as well as ultimately those that use water to drive turbines since it is the primary energy source of the water cycle on Earth. All plants use it through chemical reactions involving chlorophyll to convert water and carbon dioxide into sugar molecules which enables plants to grow and act as the pivotal base of the food chain for all organisms on Earth. This means that all Sciences (Physics, Chemistry, Biology, Astronomy, et al) have direct and minimally indirect relation to the Sun and its effects on the Earth. Today there is a new look at the Sun and its power and its importance to our needs.

Like all measures in Science, the energy given off by the Sun is just one of those quantities. From measurements a term has been developed called the Solar Constant. It is a measure of the flux (the amount of incoming solar electromagnetic radiation per unit area that is incident on a plane perpendicular to the radiation at a distance of 1 AU (astronomical unit)). This measure is all forms of energy coming from the Sun and is not just visible light.

The current measured value of the Solar Constant is 1.366 kW/m^2 on the average (note it does vary with our place in the solar system since at times of the year we are closer while other times farther away from the Sun).

This value does and has changed over time. The reasons for this are not fully known. The science of the sun is multidimensional (studying mathematical models of the interior of the Sun, helioseismology, solar sunspot activity, the mystery of the missing neutrinos, and explanations of the superheated corona, and many other areas) and far from complete. Even in 2010 it was found that the Sun's surface can have cascading fluctuations in the magnetic field causing massive solar storms – none of which were known of or had been predicted before. This also includes coronal mass ejections which even caused an electrical power outage in Canada in 1989 since it affected the Earth's magnetic field which in turn affected transformers and power lines (recall from the Electromagnet and Electric Motor Activities that changing magnetic fields generates electric currents, which happened here).

When it comes to our measure of the Solar Constant, we are only looking at visible light and none of the other electromagnetic forms of radiation. Also we have to do this from the surface of the Earth instead of in space. This means in the Activity we are leaving out many of the other energy forms, but take note that the majority of the energy is in the visible light form, so our measured approximation is a good one. Our Activity also has to consider the fact that not all of the light reaches the Earth's surface (where our measurements are made) as well as the transparency of the atmosphere (is it hazy or not due to water molecules and/or dust) and where the measurements are taking place (the angle of latitude where one is at will affect the amount of sunlight that reaches the surface since it must pass through different amounts of atmosphere during the measurements). All of this is simplified by a 'constant' factored into the Activity to compensate for these things. More precision can be found by internet research for formulae factors to incorporate as well as more precise equipment and measurements.
Note : There are two methods available in this Activity – one following the book by using a multi-meter to measure voltage & current, while the other relies on a cup of water. Your choice or do both for comparison. Explore & Enjoy! :)

Purpose : To determine the daily amount of solar energy that reaches the Earth's surface through the properties of the specific heat of water, the water's temperature change, measurement, and calculation.

Purpose : To determine an estimated value of the Solar Constant.

Materials :

- 2 Large Styrofoam cups,
- Water,
- Measuring Cup,
- Black Paint,
- Thermometer (lab quality),
- Clear Plastic Wrap,
- Flat Surface (large plate or pan or TV tray),
- Ruler,
- Paper,
- Tape,
- Clock or Stopwatch,
- Alternative Method Tools : Multimeter & Solar Cell,
- Slide Rule

Procedure :

Set Up & Needs :

1) The Radiometer (our solar energy collection device) is the two nested Styrofoam cups with the inside one having its entire interior painted black (be sure it is dry for the experiment).
2) The main need is a sunny day and the experiment is best ran in the middle of the day (between about 11 AM up to 2 PM).
3) You determine the angle to place your Radiometer at by placing the nested cups on a flat surface and elevating it so that at the time of the activity with the Sun shining on it there will be no shadow cast by the cup on the water in the cup. That is, it is directly facing the Sun.
4) For those who like numbers in this case, the angle at which the board will be to the level ground will be the same as the latitude of the location of the activity.
5) In the process of this set up determine how best to create an angled surface. A suggestion is a wood board or plastic TV tray (try not to use metal surfaces) as supported by books. Be sure to test this outside. You want the cup to hold in place, tape helps this.
6) On your angled surface place one of the Styrofoam cups and see how much water can be put in without spilling. The goal is to fill it in a way and at that angle so there is no shadow of the cup on the water's surface and the water does not spill out. (about 2/3 is good typically)

7) Measure this useful amount of water with a measuring cup. This will be the amount used in the Activity (V).

8) Note that the Thermometer will be placed in here too, so take that into consideration.

9) Always employ safety in all aspects of your Activity – do not look directly into the Sun.

Activity Procedure :

1) About 30 minutes before the Activity fill the nested cups with a measured amount of water (V) as determined from the set up above.

2) Note it is best to use room temp water neither too cold nor hot).

3) Place them on the angled surface outside facing in the general direction of where the Sun will be in 1/2 hour from now. Do not take the temperature yet. The goal is to have the water be at the same temperature as the surroundings.

4) About 5 minutes to measuring time cover the cup with the plastic wrap. At this time insert the thermometer too but still no readings.

5) It is best to insert the thermometer and place it in a manner for easy reading (to be seen from the side so as not to block the Sun striking the water's surface).

6) When the Radiometer has been in place for 5 minutes take your first temperature reading (T).

7) Record temperature readings every 2 minutes for 30 minutes (or at least until there is a temperature change of about 5°C), Δt is all of the seconds elapsed.

8) Note that you may have to slightly shift the platform with the Radiometer on it to keep it in the sunlight since the Earth is spinning and the Sun therefore 'moves' across the sky.

9) When done with the measurements, carefully remove the plastic and thermometer.

10) Use the bright-colored paint (white, yellow, or silver) to mark a line on the interior of the cup at the water line all the way around).

11) Now you can dump out the water.

12) Use scissors to cut the interior cup down to the line you have drawn.

13) Invert this angled cut surface onto some paper (it can be regular or graph paper) and trace carefully the edge of the cup as close to the inner lip as possible without distorting the cup.

14) With the traced ellipse on the paper use a ruler to measure in centimeters both the long and short axes of the ellipse (A, B).

15) Calculations to Perform :

16) Though the Slide Rule is a recommended tool, all of these calculations can be done with a regular or scientific calculator. Some scientific ones even have built-in averaging formulae. For those who like spreadsheets, the data can be typed in and the formulae then also be typed in its own cell where the formula references each of the measured variables in their respective cells.

17) Use the Ellipse formula to determine the Area of the Water's Surface. Convert this answer to m^2.

18) Calculate the amount of mass in grams, then convert to kilograms, of water used. Note the volume should have been read in mL and 1mL of water equals 1 g of mass of water.

19)

20) After determining ΔT, find the Experimental Solar Constant (Q) by plugging in all of your measured variables.
21) With the determined Solar Constant, find the percent error from the Known value of it. Compare this value to the corrected value in the table of information below.
22) Using the measured Solar Constant from the Activity, determine the amount of Power [N] (J/s or watts) reaching the Earth's surface itself.
23) Now determine the total Power [P] of the Sun's energy that reaches the Earth's orbit.
24) Since we have determined the total power (P) that reaches the sphere of Earth's radius we have indirectly found the total Power radiated by the Sun. How does th s value compare to the known expected value for the incoming solar power ?

25)   Alternative Comparison Method :
26) Use a Solar Cell (Photovoltaic Cell) connected to a Multimeter :
27) Make a Table for Voltage and Current for each of the Readings in your Trials.
28) When angled at the Sun (be sure to not be behind glass indoors), measure the Voltage (V) with the Solar Cell (most are in the neighborhood of 1.5 V),
29) Now reconnect the multimeter to measure milliamps (mA). Depending on where you live, the size of the solar cell, the overall transparency of the atmosphere this value can possibly range to over 250 mA (be sure to note the range capability of your multimeter and have it set correctly for a reading).
30) From the Voltage and Amperage readings, calculate Power : $P = V*I$
31) Measure the surface area of the solar cell actually collecting sunlight (look carefully at the solar cell to find the area actually receiving and processing the light). Convert this measurement into square-meters.
32) Like the prior calculation for the Solar Constant divide the Power determined from the electrical readings by the area of the solar cell and as before factor in your correction factor (k = 0.5 in the denominator, which in turn is really multiplying it by 2 in the numerator) for you Experimental Solar Constant Value (Q).
33) This value is good for comparative purposes and an alternative method to the aforementioned activity!

Data :

Volume of Water : _____ mL

Mass of Water : _____ kg

Ellipse : A : _____ cm
        B : _____ cm

Time (in Minutes)	Temperature (°C)
0	
2	
30 (?)	

Calculations :

Be sure to use your Slide Rule!

Constants to be used :

$1 \text{ kg} = 10^3 \text{ g}$
$1 \text{ km} = 10^3 \text{ m}$
water specific heat capacity (c) of : $1 \frac{Cal}{g*°C} = 4.186 \frac{J}{g*°C}$

    for calculations use $4.2 \frac{J}{g*°C}$

Whole Surface Area of Earth : $5.1 \times 10^8 \text{ km}^2$
Half of Earth Surface Area (SA) : $2.6 \times 10^8 \text{ km}^2$
Earth-Sun average distance (R): $1.5 \times 10^8 \text{ km}$
Correction Factor for Solar Constant Formula   (K) : 0.5
    (note this can range from 0.05 to 0.95)

Formulae :

Mass of Water : (1 mL = 1 cc, 1g H_2O is 1 mL)

    $M = D*V$

Area of Ellipse :

    $W = \frac{\pi*A*B}{4}$

A & B are the measures of the axes

Solar Constant

    $Q = \frac{m*c*\Delta T}{K*W*\Delta t}$

Power Received by the Earth on its Sun-facing Surface :

    $N = Q*(SA)$

Total Power radiated by the Sun :

    $P = 4*\pi*Q*R^2$

Percent Error : (use the values noted here)

$\%E = \frac{[\text{Experimental Value-Accepted Value}]}{\text{Accepted Value}}*100\%$

Constants to compare to :

Actual Solar Constant : 1.366 kW/m^2
Corrected Amount received at Earth for Solar Constant :
$$342 \text{ W/m}^2$$
Total amount of Solar Power reaching Earth's orbit :
$$1.74 \times 10^{17} \text{ W}$$
Energy per unit time & area received at Earth :
$$1.96 \text{ cal/(min*cm}^2)$$

Conclusion :

How well did your values match the expectations of the Solar Constant and if not, what things affected your measures?
In trying the Alternative Method, how do the values compare?

Project
Personal Slide Rule Template

MAKE A PAPER 6" SLIDE RULE

On the following two pages are two different templates than enable you to make a 6" slide rule.
Make your choice as to which you want to construct. They each have the same number and type of scales and are quite complete and useful. They have these scales for use :
C, D, C1, D1, CF, DF, C1F, A, B, S, T, ST, K, L
Things Needed : One of the Templates, Scissors, Ruler
Steps to making a slide rule :
1. In either case make 2 copies of the chosen Template.
2. For both you need something to act as a cursor – best choice is a ruler. Be sure to align it with a straight edge, such as the bottom of the paper in the case of your unfolded slide rule and be sure to fold along a straight line so as to be able to use this for the folded model.
3. In the case of the first slide rule which is the unfolded slide – this is the set of scales where they are all bunched together and in separate boxes do the following :
4. Leave the first copy alone. It acts as the stators for your slide rule and simply lie on the table
5. With the second copy, cut out the slide – the middle set of scales – so that it can be moved along between the stators as needed.
6. It is now ready to use – Use a ruler where it lays across the stators and slide perpendicular to the direction one regularly reads the paper and the bottom edge of the rule is aligned with the bottom edge of the paper (needs to be perpendicular).
7. In the case of the second slide rule which is the folded slide – this the set of scales in sets of 5 in boxes which are separated.
8. As in the first slide rule, make two copies of the template to be used.
9. With the first copy fold it so that the top and bottom set of scales are now opposite the middle set of scales. It is best to fold it so that it would align with the middle set of scales as if it were a slide and the top and bottom are the stators. – Note if there is excess paper above and below the stators so that it would interfere with the slide and their reading, cut this away in a straight a manner as possible (follow the line of the box encasing them).
10. Now fold the second template so that it fits into the sleeve and the middle set of scales faces out of the space between the top and bottom stators and is now the slide.
11. It may take some adjusting, so be patient.

12. Be sure to follow the lines of the boxes for folds as it is critical that the upper and lower stator have aligned scales.
13. With your cursor (best choice is probably a ruler as in the first case) be sure that the end edge aligns with the lines and the ruler itself is the cursor line.

Some notes for Use :
In order to use the A/B or CF/DF scales these are on the adjacent slide and stator in a doubled fashion.
Have fun and enjoy :)
Thanks :
These scales came from the web site : The International Slide Rule Museum (sliderulemuseum.com) found on the Slide Rule Reference Scales tab and set with graphics by Andrew Nikitin

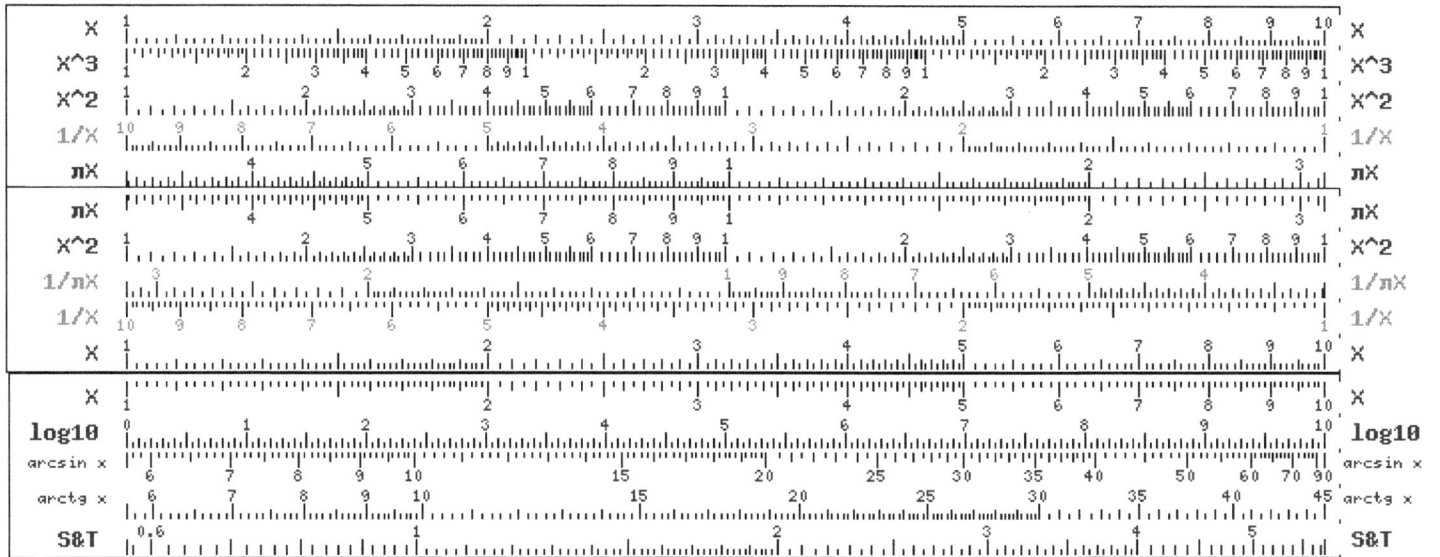

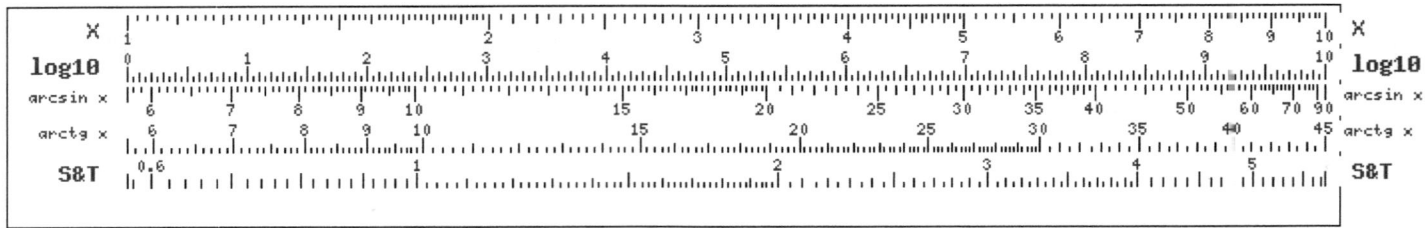

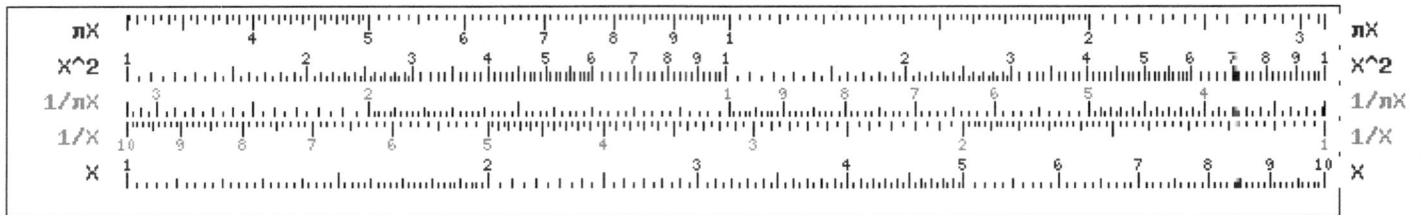

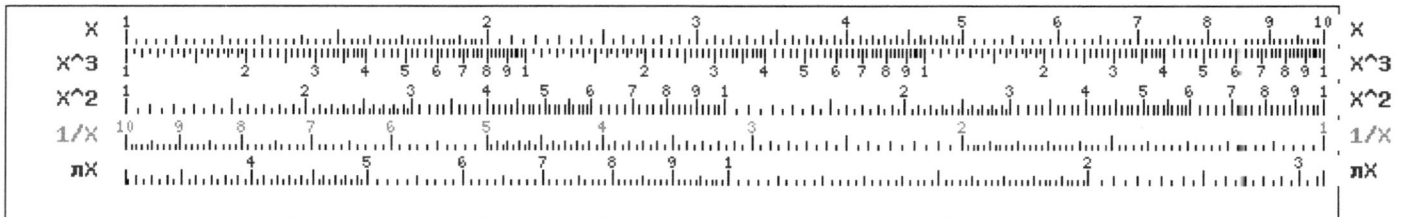